SSP 2004

SSP 2004

Proceedings of the 8th International Conference on Solid State Physics (SSP 2004), Workshop "Mössbauer Spectroscopy of Locally Heterogeneous Systems", held in Almaty, Kazakhstan, 23–26 August 2004

Edited by

K. K. KADYRZHANOV

Institute of Nuclear Physics, Almaty, Kazakhstan

and

V. S. RUSAKOV

Lomonosov Moscow State University, Moscow, Russia

Reprinted from *Hyperfine Interactions*
Volume 164, Nos. 1–4 (2005)

 Springer

A C.I.P. Catalogue record for this book is available from the Library of Congress.

ISBN 3-540-36793-4

Published by Springer
P.O. Box 990, 3300 AZ Dordrecht, The Netherlands

Sold and distributed in North, Central and South America
by Springer
101 Philip Drive, Norwell, MA 02061, U.S.A.

In all other countries, sold and distributed
by Springer
P.O. Box 990, 3300 AZ Dordrecht, The Netherlands

Printed on acid-free paper

Table of Contents

Hyperfine Interact (2005) 164: 1–2
DOI 10.1007/s10751-006-9240-6

Preface

Kairat K. Kadyrzhanov · Vyacheslav S. Rusakov

Locally inhomogeneous system is a system in which like atoms are in non-equivalent atomic positions and reveal different properties. Locally inhomogeneous systems (LIS) are, first of all, of variable composition phases, nanoscale, amorphous, multi-phase, with admixtures, defects and other systems. LIS are the most convenient model objects for studies of structure, charge, and spin atomic states, interatomic interactions, relations between matter properties and its local characteristics as well as for studies of diffusion kinetics, phase formation, crystallization and atomic ordering; all that explains considerable scientific interest in LIS. Such systems find their practical application due to wide spectrum of useful, and sometimes unique, properties that can be controlled by varying the character and degree of local inhomogeneity.

Mössbauer spectroscopy is one of the most effective methods for investigation of LIS. Local character of obtained information combined with information on cooperative phenomena makes it possible to run investigations impossible for other methods. Mössbauer spectroscopy may provide abundant information on peculiarities of macro- and microscopic states of matter including that for materials without regular structure.

June 2006.

K. K. Kadyrzhanov (✉)
Institute of Nuclear Physics, National Nuclear Center, 1 Ibragimov Str.,
050032 Almaty, Kazakhstan
e-mail: kadyrzhanov@inp.kz

V. S. Rusakov
Lomonosov Moscow State University
Leninskie gory,
119992 Moscow, Russia
e-mail: rusakov@moss.phys.msu.ru

 Springer

The 8th International Conference on Solid State Physics (SSP 2004), Workshop "Mössbauer Spectroscopy of Locally Heterogeneous Systems" was held from 23rd to 26th August 2004 in Almaty, Kazakhstan.

There were presented 35 reports by scientists from 13 countries. During four days, 37 participants discussed the following topics: *Lamellar Metal Systems, Nanocrystalline Alloys, Nanoclusters and Nanostructures, Subsurface Layers, Mechanical Alloying, Magnetic Structure and Hyperfine Interaction, Phase-Structure State of Atoms, Depth Selective Electron Mössbauer Spectroscopy, New Concepts of Data Evaluation.* Among these papers 11 are included in the present issue of "Hyperfine Interactions."

The Workshop promoted strengthening of the existing contacts and establishing new contacts between scientists from different countries with positive impact on development of key issues in Mössbauer Spectroscopy.

We are sincerely grateful to members of the International Advisory Board and to sponsors for their help and support in organization of the workshop and the conference.

We would also like to express our deepest gratitude to all the participants for the energy and efforts they put into making this workshop successful.

In additional, we would like to thank Guido Langouche for his support as the Editor-in-chief of *Hyperfine Interactions*.

We would like to extend particular thanks to the local committee member Irina Manakova for her invaluable role in keeping track of all papers.

Hyperfine Interact (2005) 164: 3–15
DOI 10.1007/s10751-006-9228-2

The Electronic and Magnetic Properties of FCC Iron Clusters in FCC 4D Metals

M. E. Elzain · A. A. Yousif · A. D. Al Rawas ·
A. M. Gismelseed · H. Widatallah · K. Bouziani ·
I. Al-Omari

Abstract The electronic and magnetic structures of small FCC iron clusters in FCC Rh, Pd and Ag were calculated using the discrete variational method as a function of cluster size and lattice relaxation. It was found that unrelaxed iron clusters, remain ferromagnetic as the cluster sizes increase, while for relaxed clusters antiferromagnetism develops as the size increases depending on the host metal. For iron in Rh the magnetic structure changes from ferromagnetic to antiferromagnetic for clusters as small as 13 Fe atoms, whereas for Fe in Ag antiferromagnetism is exhibited for clusters of 24 Fe atoms. On the hand, for Fe in Pd the transition from ferromagnetism to antiferromagnetism occurs for clusters as large as 42 Fe atoms. The difference in the magnetic trends of these Fe clusters is related to the electronic properties of the underlying metallic matrix. The local d densities of states, the magnetic moments and hyperfine parameters are calculated in the ferromagnetic and the antiferromagnetic regions. In addition, the average local moment in iron-palladium alloys is calculated and compared to experimental results.

Key words magnetic moment · hyperfine fine field · isomer shift · density of states · impurity · cluster · 4d host · iron.

1. Introduction

The development in material science has evolved to the extent of controlling sample synthesis over atomic scales. 'Engineering magnets on the atomic scale' by Kubler [1 and references therein] cites a number of examples on recent developments. Chemically different materials could be arranged in multilayers, nanodots, nanowires etc. Materials are fabricated in metastable structures not available in their corresponding

M. E. Elzain (✉) · A. A. Yousif · A. D. Al Rawas · A. M. Gismelseed · H. Widatallah ·
K. Bouziani · I. Al-Omari
Department of Physics, College of Science, Sultan Qaboos University,
Box 36, Al Khod 123, Oman
e-mail: elzain@squ.edu.om

equilibrium phase diagram. In particular, the face-centered cubic (FCC) iron has received much attention. Bulk iron in stable FCC phase exits at temperatures in the range 1,173–1,660 K. However, the FCC phase was stabilized as coherent precipitates in Cu and is reported to have antiferromagnetic (AFM) structure [2] or spin-spiral structure [3].

Herper et al. [4] using the full-potential linearized-augmented-plane-waves method (FP-LAPW) calculated the magnetic structure of bulk FCC iron and concluded that the configuration with AFM double-layer is more stable at the equilibrium lattice constant. On the other hand, Knoplfe et al. [5] using a modified augmented spherical wave method studied the energy dependence on the q vector and deduced a spin-spiral ground state.

Thin FCC iron films deposited on various surfaces were also studied and complex magnetic structures were found. A concise summary of the crystal and magnetic structures of γ-Fe on Cu surfaces is given in [6]. For the (001) surface orientation, it was found that films up to 3–4 monolayers are ferromagnetic (FM) with a tetragonally distorted FCC structure. Films with 4–11 monolayers have FCC double-layer AFM configuration, while thicker films change structure from FCC to BCC.

The work reported on thin Fe films far exceeds that reported on γ-Fe nanoparticles. There are scarce systematic studies of the variation of the crystal and magnetic structures with the size of particles embedded in other crystals [7–9]. This could be attributed to the difficulty in calculating the electronic structure due to the low symmetry resulting from introducing nanoparticles into the crystalline system. In this contribution, we present a study of the electronic and magnetic structures of coherent and relaxed γ-Fe clusters in FCC rhodium, palladium and silver as a function of the cluster size. The progress in synthesis techniques has enabled the preparation of nanoparticles of the desired size out of any material and in any selected host [10].

Iron forms solid solutions as well as ordered alloys with the 4d metals rhodium and palladium, while it is completely immiscible in silver. The rhodium-rich FCC phase exhibits competing ferro- and antiferromagnetic local spin configurations behaving as a spin glass [11]. Introduction of Fe into Pd results in a giant magnetic moment, where a large atomic moment of Fe is maintained in addition to the induced ferromagnetic polarization of the surrounding Pd atoms [12 and references therein]. On the hand, since Fe and Ag are immiscible, supersaturated out of equilibrium alloys can only be prepared by, vapor quenching, mechanical alloying or implantation [13]. At low temperatures AgFe alloys have a spin-glass phase. As the concentration of Fe varies, the alloys undergo transitions to paramagnetic or ferromagnetic phases with increasing temperature [14].

On the theoretical side the electronic and magnetic properties of Fe in the face-centered Rh, Pd and Ag were calculated using the density functional approach [7, 12, 15]. The FCC RhFe systems have received little attention and we could trace one theoretical result of Fe impurities in Rh [16], where a local moment of about 1.78 μ_B was reported. The local magnetic moment at Fe site in Pd, obtained using KKR-Green function calculation, is 3.47 μ_B. The magnetic moment at Pd neighboring sites was found to be of order 0.1 μ_B [12]. Nogueira and Petrilli [7] calculated the magnetic moments and hyperfine fields at Fe sites in Ag using the real-space linear muffin-tin orbital formalism, within atomic sphere approximation. The single Fe impurity moment of 3.07 μ_B was found to decrease as the Fe impurities start to interact. A magnetic hyperfine of -10.9 T was obtained for the isolated Fe impurity.

The objective of this work is to calculate the electronic structure and magnetic properties of Fe clusters in the structurally similar but electronically different elemental metals Rh, Pd and Ag. The study involve calculation of these properties with increasing Fe cluster size together as coherent precipitates in the elemental host metals as well as in their relaxed equilibrium FCC iron structures. We have used the FP-LAPW and supercell methodology (employing WIEN2k code) to study the properties of Fe single impurities in Rh, Pd and Ag. These are compared to the corresponding results obtained using the discrete-variational method (DVM). The latter method is then used to study the electronic and magnetic properties of Fe clusters. It is found that the results of single impurities obtained using FP-LAPW method and DVM are fairly comparable except for Rh [17]. With respect to Fe relaxed clusters it is found that antiferromagnetic coupling between the central Fe atom in the cluster and its neighboring Fe atoms, starts when the number of its Fe nearest neighbors equals 12 for Fe in Rh, whereas it starts at 42 in Pd and 24 in Ag. In the unrelaxed clusters ferromagnetic coupling is maintained at all sizes except for Fe in Rh.

This paper is organized as follows. In the following section we outline the theoretical schemes employed in the calculation. In Section 3 we present and discuss the results for Fe single impurities, the unrelaxed Fe clusters and the relaxed clusters, while the average magnetic moment per atom and per Fe atom for PdFe alloys are given in Section 4. A short summary is presented in the last section.

2. Theoretical model

The density functional theory (DFT) in its local density approximation (LDA) and its extension the generalized gradient approximation (GGA) have provided reliable and tractable frameworks for the calculation of the electronic properties of materials. Calculations start with the non-linear single particle equation

$$\left(-\frac{1}{2}\nabla^2 + V_{\mathrm{H}} + V_{\mathrm{xc}}\right)\psi = E\psi$$

where V_{H} is the usual Hartree term consisting of the average electron and electron–nucleus interaction energies and V_{xc} is the exchange-correlation potential obtained from the correlation energy E_{xc}. In the LDA, E_{xc} is a functional of the particle density, whereas in the GGA it is a functional of both density and its spatial gradient. Usually this equation is solved by, expanding the wave function ψ in terms of symmetrized basis $\{\phi_k\}$ and the equation is reduced to a set of algebraic equations, which are solved by algebraic methods. The matrix elements are evaluated as integrals over the real space. The methods employed in solving the single particle equation differ basically in the choice of the basis functions. Within each method, it is a matter of choice to use one or the other approximation for the exchange-correlation potential.

In the FP-LAPW approach, the basis functions consist of two parts. In the region of each atom spherical solutions, in addition to their first derivatives with respect to energy, where the energy is replaced by a pre-selected value, are used. In the interstitial region between the atoms, plane waves are used.

We have used the FP-LAPW method as employed in the WIEN2k code, where various forms of V_{xc} are available [18, 19]. In the calculation presented here we have

used the GGA of Perdew–Burke–Ernzerhof. Supercells of $2 \times 2 \times 2$ unit cells are used to study the electronic structure of Fe impurities in the host metals.

On the other hand, the discrete-variational method uses linear combinations of numerical atomic orbitals (LCAO) as basis functions [20]. The differential equation is reduced to the algebraic equation $(H - ES)C = 0$. The main difference between the DVM and other usual tight binding (TB) methods that employ LCAO, is in the way the matrix elements are handled. In DVM the values of the matrix elements are evaluated as sums of point-by-point values instead of being accumulated as integrals as in the usual methods. This leads to cancellation of large numbers per point, improving accuracy and reducing storage space. In the self-consistent solution of the algebraic equations, the charge density is partitioned as contributions from different atoms $\rho^{\mathrm{scc}} = \sum f_{\mathrm{nl}}^{p} |R_{\mathrm{nl}}(r_p)|^2$, where $R_{\mathrm{nl}}(r_p)$ are the radial atomic functions at the atomic position p and f_{nl}^{p} are determined by a variant of Mulliken population analysis scheme or by a least-squares error minimization procedure.

Clusters of atoms formed from the consecutive shells of a solid are used to represent the material under study. The properties of material are drawn from calculated quantities at the central atom.

In this calculation, the von Barth–Hedin approximation for the exchange-correlation potential is used. The solution of the spin-polarized equation was obtained by expanding the wave function over the valence 4d metal orbitals and the Fe impurity numerical atomic orbitals. The 3d, 4s and 4p were used for Fe, while the corresponding valence s, p and d orbitals were used for the host metals. The remaining atomic orbitals per atom were frozen. The valence orbitals were symmetrized and the resulting secular equations were diagonalized. Integration was accomplished via the diophantine sampling and Gaussian quadrature in different regions of space. Most of the calculations reported here were performed using C_{2v} point symmetry group. This allows for introducing impurity atoms one at a time in most cases. The charge density for spin σ at the central site is given by $\rho_{\sigma} = \sum_{i} n_i |\psi_{\sigma i}(0)|^2$. The contact charge density is then given by $\rho_{\uparrow} + \rho_{\downarrow}$, whereas the contact spin density is given by $\rho_{\uparrow} - \rho_{\downarrow}$. The isomer shift, in millimeters per second, at the Fe site relative to α-Fe is given by $\mathrm{IS} = \alpha[\rho - \rho_{\mathrm{Fe}}]$, where ρ_{Fe} is the contact density at Fe site in α-Fe and $\alpha = -0.24$ mm/s. The magnetic hyperfine field was assumed to result mainly from the contact spin Fermi contribution. This is split into two parts. The first part results from the polarization of core s electrons. It is proportional to the 3d moment with constant of proportionality of order 11 T/μ_{B} [21, 22]. The valence term is calculated directly from the spin density and is given by $52.4(\rho_{\uparrow} - \rho_{\downarrow})$ a.u.

3. Results and discussion

In the following subsection we first present the results for single Fe impurities in Rh, Pd and Ag using both of FP-LAPW approach and DVM. In the following subsection we present the results for clusters of γ-Fe in the same metals using the DVM only.

3.1. Single iron impurities

We have calculated the electronic and magnetic properties of γ-Fe at its equilibrium lattice constant of 6.8 a.u. and at the expanded lattice constants of 7.18, 7.35 and

 Springer

Table I The local magnetic moment (μ_{tot}), the isomer shift (IS) relative to α-Fe and the hyperfine field (B_{hf}) at Fe sites in the expanded FCC structure with lattice constants (a) corresponding to those of Rh, Pd and Ag obtained using method (1) DVM and (2) FP-LAPW formalisms

a (a.u.)	7.18		7.35		7.73	
Method	1	2	1	2	1	2
μ_{tot} (μ_B)	1.98	2.73	2.75	2.86	2.88	3.01
IS (mm/s)	0.16	0.15	0.24	0.23	0.39	0.34
B_{hf} (T)	−35	−39	−41	−38	−42	−37

Table II The local magnetic moment (μ_{tot}), the isomer shift (IS) relative to α-Fe and the hyperfine field (B_{hf}) at Fe site in the FCC metals Rh, Pd and Ag obtained using method (1) DVM and (2) FP-LAPW formalisms

System	RhFe		PdFe		AgFe	
Method	1	2	1	2	1	2
μ_{tot} (μ_B)	2.22	2.94	3.50	3.43	2.93	3.07
IS (mm/s)	0.02	0.12	0.23	0.23	0.49	0.42
B_{hf} (T)	−19	−14	−18	−18	−16	−11

7.73 a.u. corresponding to those of Rh, Pd and Ag, respectively. In addition, the properties of BCC iron were also calculated using both methods. The local magnetic moments at Fe site in BCC were found to be 2.27 and 2.30 μ_B from FP-LAPW and DVM, respectively, whereas the respective magnitudes of magnetic hyperfine fields (B_{hf}) are 32 and 39 T. The results from DVM are larger due to the finite cluster size effects. The contact charge densities at Fe BCC sites were also computed. These are needed for the calculation of the Mössbauer isomer shifts (IS). Table I shows the local magnetic moments, IS and B_{hf} at Fe sites in the expanded FCC iron.

In Table I we observe that the values obtained using the two methods are in agreement for $a = 7.35$ and $a = 7.73$ a.u. However, for $a = 7.18$ a.u. the DMV gives relatively smaller values. We recall that the DMV gives an AFM solution for $a = 7.0$ a.u., while a FM solution is obtained by FP-LAPW. This could be attributed to the number of atoms used in each calculation, where 55 atoms are used in DVM, while a one-atom primitive cell is used in FP-LAPW. Indeed the result obtained by, DVM using a 19-atom cluster gives a larger moment.

The isomer shifts, relative to α-Fe, obtained by the two methods are in reasonable agreement. The contact hyperfine fields obtained using the DVM are in general large because of the large contribution resulting from the valence component due to the cluster size effect. Apart from these larger DVM values, the trends of the hyperfine fields are satisfactorily in agreement.

Table II shows the local magnetic moment, IS and B_{hf} at Fe sites in Rh, Pd and Ag. The results for Fe in Pd and to a reasonable degree for Fe in Ag are comparable and agree with experimental results. However, the results for Fe in Rh are in disparity exhibiting the trends discussed above for the expanded γ-Fe with $a = 7.18$ a.u. (the lattice constant of Rh). We note that Fe magnetic moment calculated using DVM is in agreement with the experimental results [23] and larger than that reported in [16]. In Figure 1 the partial d densities of states (DOS) at Fe site in Rh, Pd and Ag are presented. The results are reasonably comparable when taking into account that the

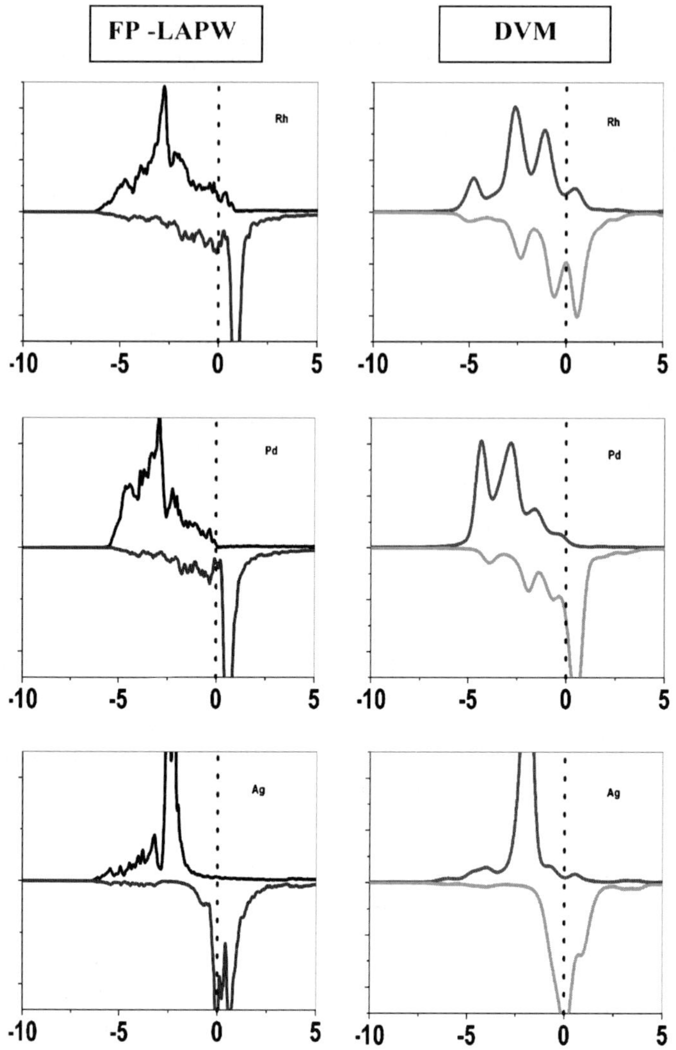

FP -LAPW	DVM

Figure 1 The d local density of states of Fe site Rh, Pd and Ag obtained using FP-LAPW method (*left*) and DVM (*right*). *Dotted vertical lines* indicate the position if the Fermi level. Energies are in eV.

broadening used in extracting densities in FP-LAPW method is different from that of DVM.

When comparing the results shown in Tables I and II, we observe that the values of the local moment at Fe in the expanded γ-Fe with $\alpha = 7.73$ a.u. and in Ag are close, while the hyperfine fields are very different. The difference between the isomer shifts is negligible. This indicates that the d–d interaction is weak in this expanded γ-Fe as well as in Ag. On other hand the s–d interaction is strong and in the case of the expanded γ-Fe it leads to a negative contribution to the hyperfine field and positive contribution to Fe in Ag. The total charge density is slightly affected.

The electronic d–d interactions of Fe in Rh and Fe in Pd are strong leading to changes in the magnetic moments and to charge transfer. We propose that the

 Springer

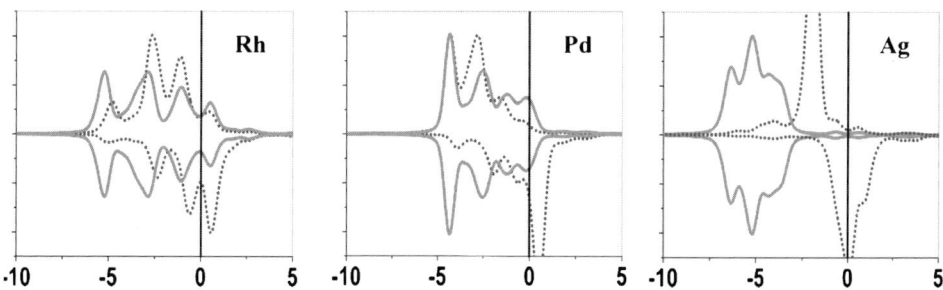

Figure 2 The partial d DOS at host metal sites (*solid curve*) and at Fe impurity in the corresponding host metal (*dotted curve*) for Rh, Pd, Ag. The *vertical lines* indicate the position of the Fermi level. Energies are in eV.

d electrons of Pd interact more with Fe majority electrons and less with the minority electrons leading to the observed large increase in the Fe magnetic moment accompanied with a loss of about 0.2 d-electrons from Fe to Pd. This also induces positive moments at the Pd sites. On the other hand the Rh d-electrons interact comparably with Fe majority and minority electrons giving the same d-moment as that for expanded γ-Fe. The observed increase in Fe local moment is due to sp contribution. Indeed, the components of local moments at Fe site in Rh are μ_d=2.04 μ_B and $\mu_{sp} = 0.18$ μ_B, whereas the corresponding moments for the expanded γ-Fe are $\mu_d = 2.00$ μ_B and $\mu_{sp} = -0.02$ μ_B, respectively. Consequently, negative moments are induced on Rh atoms since the Fe minority *d*-states extend more into the Rh sites. The hyperfine field is reduced because the induced moments at Rh and Pd sites are small and not enough to generate negative contributions to the hyperfine field through s–d interaction. The isomer shifts do not change very much except for Fe in Rh as compared to the corresponding values of expanded γ-Fe with Rh lattice constant.

This picture is confirmed when considering the d local DOS and induced polarization in the host metal. Figure 2 shows the local d DOS at Fe and host metal sites. While the DOS at Ag lies well below the Fermi level, the corresponding densities for Rh and Pd overlap that of Fe. The Fermi level in Pd lies just above the leading peak, whereas it lies well below the peak for Rh. This trend is attributed to the difference in the polarization of these host metals [24]. There is no induced magnetic moment at Ag site except for small negative moment due to the s electrons. On the other hand, positive d moments accompanied by negative sp moments are induced at the Pd neighboring sites, while both d and sp moments induced at Pd sites in the outer shells are positive. The induced moments at the Rh sites were found to be both negative.

Armed with the confidence gained from the reasonably good agreement between the results obtained for Fe impurities using the FP-LAPW approach and the DVM, we venture into considering application of DVM to study the properties of Fe clusters in Rh, Pd and Ag.

3.2. FCC iron clusters

The results presented below are for Fe atoms as they are added consecutively to form the FCC clusters, which are coherent with the host lattice as well as for clusters with the equilibrium lattice constant of γ-Fe. Calculations are performed using the DVM.

 Springer

Figure 3 The local magnetic moment (*open circles*), the contact charge density (*full box*) and the hyperfine field (*full circle*) at the central Fe site, scaled by the corresponding single Fe impurity in Ag quantities, *versus* the iron cluster size *N*. The iron clusters are coherent with the underlying silver FCC lattice. *Lines* are for eye guidance.

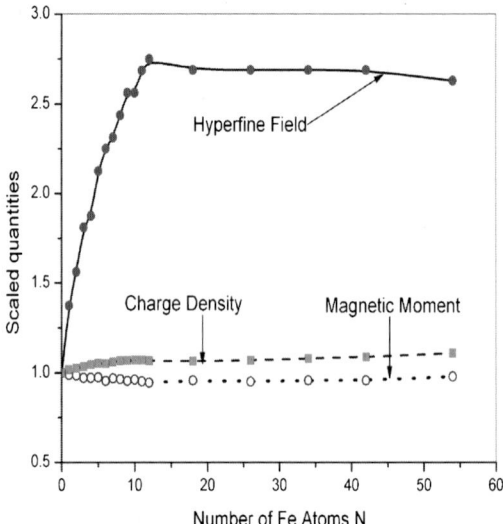

The clusters grow from the center outwards. The shell of first neighbors is filled first and then followed by the shell of second neighbors and so on.

We consider Fe in Ag first because it is simpler. When iron grows coherently within the silver FCC lattice it forms a FM phase. With Fe atoms in this phase, we have calculated the local properties at the central atom for Fe clusters extending from one Fe atom to 55 Fe atoms. The contact charge density and local magnetic moment at central Fe atom remain almost constant with increasing number of Fe atoms, while the local magnetic hyperfine field increases steadily with increasing number of surrounding Fe atoms, reaching saturation at 12 Fe atoms (i.e. at filled nearest neighbor shell). These trends are illustrated in Figure 3, where the three quantities are scaled by the corresponding single Fe atom values. Consequently, it is expected that in a system of FM iron nanoparticles in Ag, the magnetization per Fe atom and the Mössbauer isomer shift remain constant as Fe content or Fe particle size change. On the other hand, the average and the distributions of the contact hyperfine field vary with particle size since the Fe atoms in the nanoparticle have different local environment. However, the observation that magnetic fields saturate for filled nearest neighbor shell configuration, simplify the deduction of fields in larger clusters. For a cluster of N iron atoms with shells filled consecutively from the center, we calculated the number of Fe atoms, n_j, with j Fe atoms in its neighboring shell. The hyperfine field distribution, $P_N(H)$, is then determined using

$$P_N(H) = \sum_{j=0}^{12} \frac{n_j}{N} \delta(H - B_j)$$

where B_j, is the calculated field at the central Fe atom with j Fe neighbors. For example, we show the in Figure 4 the field distribution for clusters with 13, 141 and 16,757 Fe atoms. In the 13 Fe atom clusters the distribution is peaked around a value corresponding to the surface atoms with tails extending into values corresponding to the bulk. On the other hand, the distribution for the 16,757 atoms cluster is peaked around the bulk hyperfine field with tails extending into the lower fields region. The

 Springer

Figure 4 The contact hyperfine field distribution $P_N(H)$ for 13 (*dash-dotted*), 141(*continuous*) and 16,757 (*dotted*) iron atom clusters with FCC crystal structure coherent with that of silver.

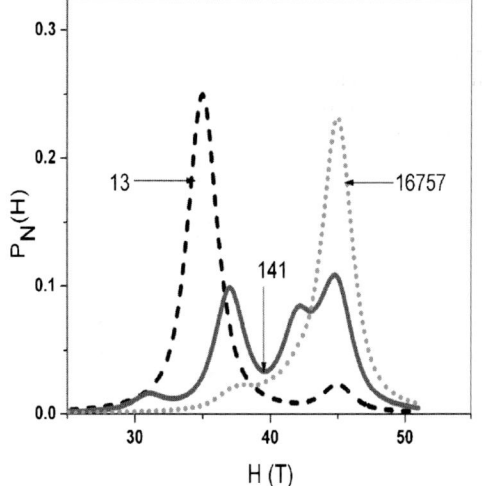

Figure 5 The local magnetic moment (*open circles*), the contact charge density (*full box*) and the hyperfine field (*full circle*) at the central Fe cite, scaled by the corresponding single Fe impurity in Ag quantities, *versus* the iron cluster size N. The iron clusters form a relaxed FCC structure in the underlying silver lattice. *Lines* are for eye guidance.

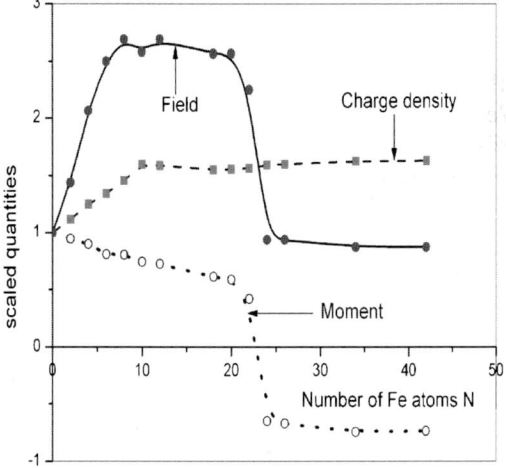

field distribution for the 141 atoms cluster spreads over a wide range indicating the presence of various local environments. Of course some Ag atoms may be distributed inside the Fe clusters and this will lead to additional structures in the field distribution. We note that Morales et al. in [13] reported a distribution of hyperfine fields at 4.2 K with three peaks.

When the lattice constant of the FCC iron clusters is reduced from Ag lattice constant of 4.09 to 3.60 Å the trends of the local properties undergo various changes. In Figure 5 we show the trends of the local magnetic moment, the contact magnetic hyperfine field and the contact charge density at the central Fe site *versus* the number of iron atoms N. The values are scaled relative the corresponding single Fe impurity quantities. The local moment decreases with increasing N and for $N > 23$, the central Fe atoms couples AFM to its neighbors. The magnetic hyperfine field increases initially reaching saturation in a similar manner to the expanded lattice.

Figure 6 The local magnetic moment (*open circles*), the contact charge density (*full box*) and the hyperfine field (*full circle*) at the central Fe site, scaled by the corresponding single Fe impurity in Pd quantities, *versus* the iron cluster size N. The iron clusters are coherent with the underlying palladium FCC lattice. *Lines* are for eye guidance.

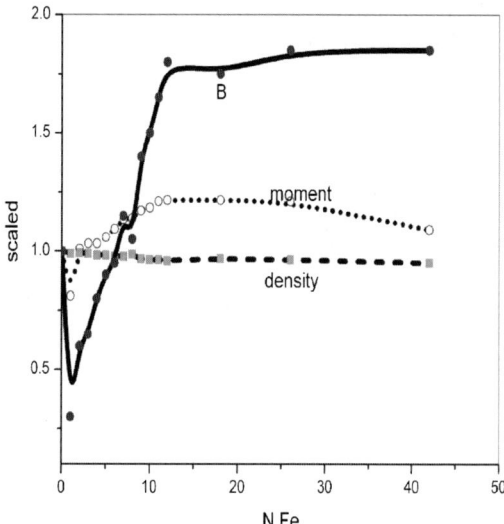

Figure 7 The local magnetic moment (*open circles*), the contact charge density (*full box*) and the hyperfine field (*full circle*) at the central Fe site, scaled by the corresponding single Fe impurity in Rh quantities, *versus* the iron cluster size N. The iron clusters are coherent with the underlying rhodium FCC lattice. *Lines* are for eye guidance.

However for $N > 23$, where the central Fe atoms couple AFM to its surroundings, the hyperfine field drops to small values. The contact charge density increases initially with increasing N and then remains constant.

The clusters of Fe in Pd and Rh do not show the simple trends exhibited by Fe in Ag. In Figure 6 we show the scaled magnetic moments, hyperfine fields and contact charge density at Fe in Pd *versus* the cluster size. While the contact charge density remains almost constant, the magnetic moment decreases with increasing cluster size and saturates after the first shell is filled with Fe atoms. On the other hand, the magnetic hyperfine field increases sharply with increasing number of Fe in the first shell until it is full and then increasing gradually afterwards. In fact the hyperfine field

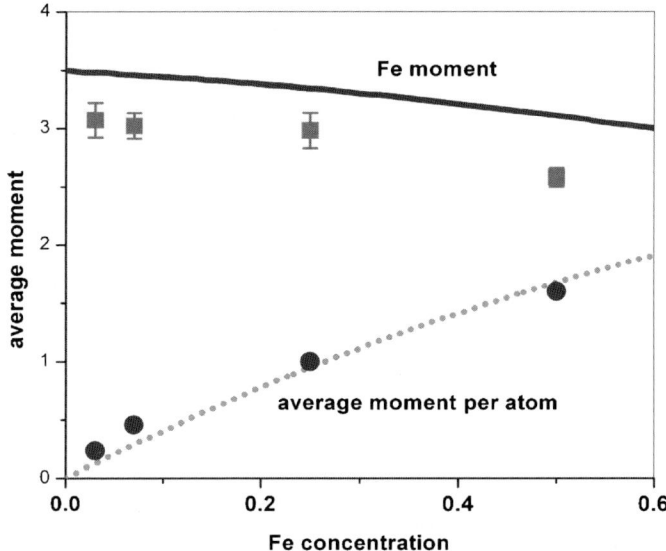

Figure 8 The calculated average magnetic moment per atom (*dotted curve*) and per Fe atom (*continuous curve*) *versus* the Fe concentration in PdFe alloys. The *filled circles* and *boxes* are the experimental results of Cable et al. [25].

was found to decrease in magnitude with increasing number of Fe atoms in the second shell for fixed numbers in the first shell. For the relaxed clusters the ferromagnetic coupling between the central Fe atom and its Fe neighbors is maintained up to about 40 Fe atoms.

For Fe in Rh the contact charge density remains constant with increasing cluster size. However, the magnetic moment and hyperfine field exhibit more complex trends (Figure 7). For less than complete shell of first neighbors the magnetic moment and hyperfine field reflect oscillatory features. The host Rh atoms were found to show AFM coupling in general. In addition, the central atoms were found to couple antiferromagnetically with Fe atoms in the second shell of neighbors.

4. Iron-palladium alloys

As a further test to the accuracy of the results obtained using the DVM we consider the average magnetic moments per atom and per iron atom for PdFe alloys and compare that to the experimental results of Cable et al. [25]. The local magnetic moments at Fe and Pd sites in FCC structure were calculated for a number of configurations to span various environments composed of the first and second neighboring shells of the central Fe and Pd atoms. We assume that the Fe atoms in PdFe alloys are distributed randomly according to a binomial distribution. The probability of a configuration of N iron atoms in the first shell and M iron atoms in the second shell of neighbors is given by

$$p_{NM} = \frac{12!}{(12-N)!N!}\frac{6!}{(6-M)!M!}c^{N+M}(1-c)^{18-N-M}$$

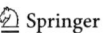 Springer

where c is the Fe concentration. The average moment contributed by atom X ($X =$ Fe or Pd) is

$$\bar{\mu}_X = \sum_{N,M} p_{NM}\mu_X(N, M)$$

Hence the average moment per atom is obtained as

$$\bar{\mu} = c\bar{\mu}_{Fe} + (1 - c)\bar{\mu}_{Pd}$$

Figure 8 shows the calculated average moments and the corresponding experimental results [25]. The agreement between the two results is reasonably good in particular for the average moment per atom.

5. Summary

The electronic and magnetic structures of Fe impurities in FCC metals Rh, Pd and Ag were calculated using the FP-LAPW method and DVM. The results of both methods were found to be comparable and in agreement with the experimental results for Fe in Pd and Ag. On the other hand, disagreement was found for Fe in Rh results.

The DVM was used to study the magnetic structures of Fe clusters in the FCC metals Rh, Pd and Ag. The trends of the coherent Fe clusters in Ag are simple and were used to predict the properties of larger particles. For relaxed clusters, the central Fe atom was found to couple antiferromagnetically to its Fe neighbors for clusters of size larger than 23 Fe atoms. The clusters in Pd retain ferromagnetism in the coherent structure and for relaxed clusters of size up to 42 Fe atoms. The magnetic moment at Fe in Pd was found to decrease with increasing number of Fe neighbors, whereas the hyperfine field was found to vary with changes in the number of Fe atoms in the second shell of neighbors. The trends of Fe clusters in Rh were found to be complex due to dominant antiferromagnetic couplings between Fe and Rh atoms as well as between Rh atoms.

The average magnetic moments of PdFe alloys *versus* Fe concentration were calculated and found to agree with experimental results.

References

1. Kubler, J.: J. Phys., Condens. Matter **15**, V21 (2003)
2. Abraham, S.C., Guttman, L., Kasper, J.S.: Phys. Rev. **127**, 2052 (1962)
3. Tsunoda, Y.: J. Phys., Condens. Matter **1**, 10427 (1989); Tsunoda, Y., Nishioka, Y., Nicklow, M., J. Magn. Magn. Mater. **128**, 133 (1993)
4. Herper, H.C., Hoffmann, E., Entel, P.: Phys. Rev. B **60**, 3839 (1999)
5. Knoplfe, K., Sanratskii, L.M., Kubler, J.: Phys. Rev. B **62**, 5564 (2001)
6. Spisak, D., Hafner, J.: J. Magn. Magn. Mater. **272–276**, 1184 (2004)
7. Nogueira, R., Petrilli, H.: Phys. Rev. B **60**, 4120 (1999)
8. Li, Z., Hashi, Y., Kawazoe, Y.: J. Magn. Magn. Mater. **167**, 123 (1997)
9. Ellis, D.E., Guo, J., Lam, D.J.: Rev. Solid State Sci. **5**, 287 (1991)
10. Martin, J.I., Nogues, J., Liu, K., Vicent, J.L., Schuller, I.K.: J. Magn. Magn. Mater. **256**, 449 (2003); Fassbender, J., Ravelosona, D., Samson, Y.: J. Phys., D. Appl. Phys. **37**, R179 (2004)
11. Parfenova, V.P., Delyagin, N.N., Erzinkyan, A.L., Reyman, S.I.: Phys. Status Solidi, B **214**, R1 (1999); Parfenova, V.P., Erzinkyan, A.L., Delyagin, N.N., Reyman, S.I.: Phys. Status Solidi, B **228**, 731 (2001)

12. Gubanov, V.A., Liechtenstein, A.I., Postnikov, A.V.: Magnetism and Electronic Structure of Crystals, Springer Series in Solid-State Sciences 98, p. 125. Springer, Berlin Heidelberg New York (1992)
13. Ma, E., He, J.-H., Schilling, P.J.: Phys. Rev. B **55**, 5542 (1997); Morales, M.A., Passamani, E.C., Baggio-Saitovitch, E.: Phys. Rev. B **66**, 144422 (2002)
14. Manns, V., Scholz, B., Keune, W., Schletz, K.P., Braun, M., Wassermann, E.F.: J. Physique, Colloque **C8**(suppl. 12) 1149 (1988)
15. Moon, H., Kim, W., Oh, S., Park, J., Park, J.G., Cho, E., Lee, J., Ri, H.: J. Korean Phys. Soc. **36**, 49 (2000); Shi, Y., Qian, D., Dong, G., Wang, D.: Phys. Rev., B **65**, 172410 (2002)
16. Hoshino, T., Shimizu, A., Zeller, R., Dederichs, P.H.: Phys. Rev. B **53**, 5247 (1996)
17. Elzain, M.E., Al Rawas, A.D., Yousif, A.A., Gismelseed, A.M., Rais, A., Al Omari, I., Widatallah, H.: Phys. Status Solidi, C **1**, 1796 (2004)
18. Blaha, P., Schwarz, K., Madsen, G.K.H., Kvasnicka, D., Luitz, J.: WIEN2k, An Augmented Plane Wave + Local Orbitals Program for Calculating Crystal Properties (Karlheinz Schwarz, Tech. Universitat Wien, Austria), 2001. ISBN 3-9501031-1-2
19. Cottenier, S.: Density Functional Theory and the family of (L)APW-methods; a step-by-step introduction (Institute voor Kern-en Stralingsfysica, K. U. Leuven, Belgium), 2002, ISBN 90-807215-1-4 (to be found at http://www.wien2k.at/reg_user/textbooks)
20. Averil, F.W., Ellis, D.E.: J. Chem. Phys. **59**, 6412 (1973)
21. Elzain, M.E., Ellis, D.E., Guenzberger, D.: Phys. Rev. B **34**, 1430 (1986)
22. Battocletti, M., Ebert, H.: Phys. Rev. B **53**, 9776 (1996)
23. Clogston, A.M., Matthias, B.T., Peter, M., Williams, H.J., Corenzwit, E., Sherwood, R.C.: Phys. Rev. **125**, 541 (1962)
24. Moruzzi, V.I., Marcus, P.M.: Phys. Rev. B **39**, 471 (1989)
25. Cable, J.W., Wollan, E.O., Koehler, W.C.: Phys. Rev. **138**, A755 (1965)

Hyperfine Interact (2005) 164: 17–26
DOI 10.1007/s10751-006-9229-1

Magnetic Properties of Co and Ni Substituted ε-Fe$_3$N Nanoparticles

N. S. Gajbhiye · R. S. Ningthoujam ·
Sayan Bhattacharyya

Abstract In the present work, the nanostructured pseudo-binary ε-Fe$_{3-x}$Co$_x$N and ε-Fe$_{3-x}$Ni$_x$N ($x = 0.0$–0.8) systems (10–20 nm) are studied in detail for their intrinsic magnetic properties. These systems are indexed based on the hexagonal ε-Fe$_3$N phase with the space group P6$_3$/mmc. However, for the cobalt containing samples, $x = 0.4$–0.8, a mixture of hexagonal ε-Fe$_{3-x}$Co$_x$N and bcc α-Fe phases are formed and for the nickel containing systems, $x = 0.5$–0.8, small amount of cubic γ'-Fe$_{4-y}$Ni$_y$N is precipitated and these facts are supported by X-ray diffraction and ^{57}Fe Mössbauer spectroscopy studies. Mössbauer studies also confirm the random occupation of the substituent elements and the superparamagnetic nature of these particles. The intrinsic magnetic properties (σ_s, T_c) are observed to vary with the Co and Ni contents. The low temperature magnetic properties are dominated by exchange bias phenomena and spin-glass like ordering, for these compositions.

Key words nitrides · nanocrystalline · superparamagnetism · spin-glass like nature.

1. Introduction

The nitrides of d-block transition metals are commercially important due to their hardness, electrical, magnetic and mechanical properties. Iron nitrides have been extensively studied in the literature because of their interesting magnetic properties [1, 2]. Many theoretical studies are available on the electronic structure of the iron

N. S. Gajbhiye (✉) · R. S. Ningthoujam · S. Bhattacharyya
Department of Chemistry, Indian Institute of Technology,
Kanpur 208 016, U.P., India
e-mail: nsg@iitk.ac.in

N. S. Gajbhiye
Institute of Nanotechnology, Forschungszentrum Karlsruhe,
P.O. Box 3640, 76021 Karlsruhe, Germany

nitrides in order to explain the origin of their magnetism [3]. ε-Fe$_3$N is a soft ferro-magnetic material and has the potential for the applications in magnetic recording media. Among the various iron nitrides known till today, the ε-phase covers the widest range of nitrogen concentration in the iron–nitrogen phase diagram [4]. Upon substitution of other ferromagnetic elements into ε-Fe$_3$N, it is expected to change the magnetic properties significantly. Although substitutions of other elements into γ'-Fe$_4$N has been reported in the literature [5–7], not much has been studied for the substitutions in ε-Fe$_3$N. However, Takahashi et al. studied the substitution of Co into ε-Fe$_3$N [8, 9]. Moreover ε-Fe$_3$N in the nanoparticle form will have magnetic properties quite different from the bulk, since the various intrinsic magnetic properties (σ_s, T_c, H_c, H_f) are found to depend on the particle size. Magnetic properties are affected due to the surface oxidation of ε-Fe$_3$N nanoparticles, which results into spin canting at the surface.

In this paper, we report the substitution of Co and Ni into ε-Fe$_3$N nanoparticles, to give the compositions ε-Fe$_{3-x}$Co$_x$N ($0.0 \leq x \leq 0.8$) and ε-Fe$_{3-x}$Ni$_x$N ($x = 0.0$–0.8) and the structural and magnetic properties are investigated. Room temperature ^{57}Fe Mössbauer spectroscopy studies are correlated to the observed magnetic properties for both these systems.

2. Experimental section

The ε-Fe$_{3-x}$Co$_x$N ($0.0 \leq x \leq 0.8$) and ε-Fe$_{3-x}$Ni$_x$N ($x = 0.0$–0.8) systems are synthesized using oxalate and citrate precursor routes, respectively. The precursors are decomposed in air atmosphere with subsequent nitridation in NH$_3$ (g) atmosphere at varied temperatures and time, to get the desired nanostructured phases. For the ε-Fe$_{3-x}$Co$_x$N ($0.0 \leq x \leq 0.8$) system, the nitridation reaction is carried out in the temperature range (773–823 K) for 4 h, while for ε-Fe$_{3-x}$Ni$_x$N ($x = 0.0$–0.8), the Fe–Ni-oxides are nitrided in the range (673–823 K) for 12 h, to obtain the desired compositions. The stoichiometry of the final products is confirmed by chemical analyses of iron, cobalt, nickel and nitrogen amounts.

These nitride nanoparticles are characterized for their structure by using a Rich–Seifert X-ray diffractometer, model Isodebyeflex 2002 with CuK$_\alpha$ radiation and Ni filter. The crystallite size is determined using Scherrer's relationship, $D = K\lambda/B\cos\theta$, where D is the diameter in Ångström, K is a constant (shape factor), B is the half-maximum line width and λ is the wavelength of the X-rays used. Room temperature ^{57}Fe Mössbauer spectroscopy experiments are performed using a ^{57}Co(Rh) source and the spectra are analyzed by Mössbauer software PC-MOS obtained from CMTE-FAST Electronik, Germany. The Mössbauer data are presented with respect to natural iron. The magnetic measurements are performed with vibrating sample magnetometer (Par-150A), which can provide a maximum magnetic field of 1.1 T and Superconducting Quantum Interference Device (SQUID, M/S Quantum Design USA) in the temperature range 5–300 K.

3. Results and discussion

Analysis of the XRD patterns, with the help of Rietveld analysis program, for the pseudo-binary ε-Fe$_{3-x}$Co$_x$N ($0.0 \leq x \leq 0.8$) and ε-Fe$_{3-x}$Ni$_x$N ($x = 0.0$–0.8) systems

Figure 1 X-ray lattice
parameters (a, c, c/a) as a
function of x, where x is Co/Ni
substituted in ε-Fe$_3$N.

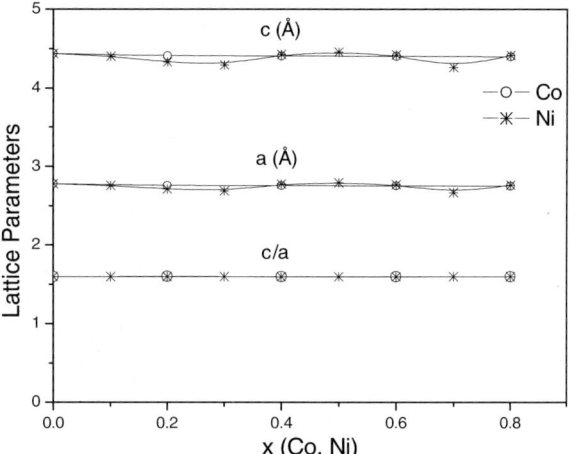

confirm the formation of ε-Fe$_3$N hexagonal structure for the compositions of $x = 0.1$–0.2 and $x = 0.1$–0.4, respectively. For $x = 0.6$–0.8 of ε-Fe$_{3-x}$Co$_x$N system, a mixture of hexagonal phase and bcc α-Fe phases are formed while for ε-Fe$_{3-x}$Ni$_x$N system, $x = 0.5$–0.8, a small amount of fcc γ'-Fe$_{4-y}$Ni$_y$N is also precipitated along with the hexagonal phase. Since for $x = 0.5, 0.6, 0.7$ and 0.8, the Ni percentages are 16.67, 20, 23.33% and 26.67%, respectively, the corresponding values of Ni concentration (y) in γ'-Fe$_{4-y}$Ni$_y$N phase are 0.7, 0.8, 0.9 and 1.1, respectively. The XRD pattern of the hexagonal phase for both the systems, correspond exactly to the ε-Fe$_3$N system. The XRD lines are relatively broad and indicate the fine particle nature of the nitride materials. The average crystallite size for both the systems is in the range 10–20 nm. The XRD patterns are refined with Rietveld refinement program with the space group P6$_3$/mmc. In these substituted ε-Fe$_3$N phases, Co and Ni substitute the Fe atoms randomly. The metal atoms occupy [1/3, 2/3, 1/4] and [2/3, 1/3, 3/4] positions while the N atoms occupy [0, 0, 0] and [1/3, 2/3, 1/2] positions. For ε-Fe$_{3-x}$Co$_x$N system, the lattice parameter 'a' decreases from 2.759(9) to 2.751(3) Å and 'c' does not show any trend and varies from 4.405(9) to 4.396(7) Å, for $x = 0.0$–0.8. For the Ni substituted system, they do not show any particular trend with the increasing Ni concentration ($x = 0.0$–0.8), with 'a' varying from 2.779(1) to 2.668(6) Å and 'c' varying from 4.438(2) to 4.259(2) Å as shown in Figure 1. c/a ratio for both these systems are nearly same and almost stays constant over the whole composition range. The electron microscope studies by TEM reveal the particles to be nearly spherical with the presence of agglomeration (Figure 2). The particle size in the Co substituted system is in the range 10–25 nm while for the Ni substituted system, average particle size is 30 nm.

Tables 1 and 2 present the room temperature Mössbauer spectral parameters for the ε-Fe$_{3-x}$Co$_x$N and ε-Fe$_{3-x}$Ni$_x$N systems, respectively. For the Mössbauer spectra for ε-Fe$_{3-x}$Co$_x$N system, $x = 0.1$–0.2, and for the ε-Fe$_{3-x}$Ni$_x$N system, $x = 0.1$–0.4, only superparamagnetic doublet is observed. However, for the higher substitutions in both the systems, a mixture of superparamagnetic doublet and ferromagnetic sextets are present, which is due to the precipitation of bcc α-Fe for $x = 0.4$–0.8 in ε-Fe$_{3-x}$Co$_x$N system and precipitation of fcc γ'-Fe$_{4-y}$Ni$_y$N phase in the ε-Fe$_{3-x}$Ni$_x$N

Figure 2 TEM micrographs of
a ε-Fe$_{2.4}$Co$_{0.6}$N and
b ε-Fe$_{2.6}$Ni$_{0.4}$N.

system, for $x = 0.5$–0.8. The superparamagnetic doublet observed in ε-Fe$_{3-x}$Co$_x$N and ε-Fe$_{3-x}$Ni$_x$N systems are explained from ε-Fe$_3$N structure, where Fe atoms are in the ε-Fe$_3$N lattice site and may be replaced randomly by the added Co and Ni atoms and N atoms occupy regular positions. Since particle size is less than the critical superparamagnetic particle size at 300 K, magnetization direction fluctuates spontaneously. Similar superparamagnetic doublet is observed in ultrafine pure ε-Fe$_3$N particles prepared by chemical methods [10].

In ε-Fe$_{3-x}$Co$_x$N system the isomer shift (δ) value is almost the same for all the compositions (0.45 mm/s) and for the ε-Fe$_{3-x}$Ni$_x$N system, the δ values for the superparamagnetic doublet range between 0.43 and 0.45 mm/s for $x = 0.0$–0.4 [11], but decrease to 0.39 mm/s for the higher Ni concentrations. These values are higher than the value of 0.33 mm/s for the Fe sites reported for γ'-Fe$_4$N [12]. This indicates that in these systems, Fe has smaller s-electron density or greater p- and d-electron density compared to that of γ'-Fe$_4$N and hence there is an interaction of N and Fe atoms in ε-Fe$_3$N system. In the Fermi-level (E_f), the number of electrons in ε-Fe$_3$N system is due to the mixing of d-electrons of Fe and p-electrons of N. The density of states arises from the unbalanced spins between Fe 3d-orbital and N 2p-orbital

 Springer

Table 1 ^{57}Fe Mössbauer resonance parameters for ε-Fe$_{3-x}$Co$_x$N $(0.0 \leq x \leq 0.8)$

Compositon Co (x)	Hyperfine field ΔH_f (T)	Isomer shift δ (mm/s)	Quadrupole splitting Δ (mm/s)	Line widths Γ (mm/s)
0.2				
(doublet)	–	0.44	0.12	0.47
0.4				
(sextet)	40.14	0.0	−0.05	0.46
(doublet)		0.45	0.15	0.42
0.6				
(sextet)	39.68	0.0	−0.06	0.47
(doublet)		0.46	0.16	0.86
0.8				
(sextet)	39.74	0.0	−0.04	0.63
(doublet)		0.45	0.16	0.60

electrons, which results in a partial covalent-character between Fe and N bonds. The isomer shifts also support the covalent character [10]. No major change in the quadrupole splitting values in doublet is observed, as the Co and Ni concentrations are increased and this indicates the random distribution of Co and Ni atoms at Fe sites of ε-Fe$_3$N lattice.

The magnetic sextet for ε-Fe$_{3-x}$Co$_x$N $(x = 0.4$–$0.8)$ system is observed due to the presence of α-Fe phase, which is confirmed from the zero isomer shift values [13]. For ε-Fe$_{3-x}$Ni$_x$N system $(x = 0.5$–$0.8)$, the two sextets are due to two different sites, the corner position (Fe^c) and the fcc position (Fe^f) of the fcc lattice of the γ'-Fe$_{4-y}$Ni$_y$N phase. For the ε-Fe$_{3-x}$Co$_x$N system, the hyperfine field (ΔH_i) value of 40 T is in good agreement with that of the nanostructured metallic Fe particles [2].

For the compositions $x = 0.5$–0.8, for Ni substituted samples, the sextet having the larger ΔH_i corresponds to the corner Fe atoms and the other sextet with smaller ΔH_i corresponds to the face centered Fe atoms in the γ'-Fe$_{4-y}$Ni$_y$N phase. In the γ'-Fe$_{4-y}$Ni$_y$N phase, the added Ni atoms preferentially substitutes the corner Fe positions in the γ'-Fe$_4$N lattice, which is justified from the fact that the sextet corresponding to the corner sites of γ'-Fe$_{4-y}$Ni$_y$N phase have lower intensities than the sextet for the face centered Fe sites. The percentage of Ni atoms occupying the corner positions in γ'-Fe$_{4-y}$Ni$_y$N phase increases from 78.82% for $y = 0.7$ to 88.21% for $y = 0.8$ and then decreases to 69.09% for $y = 0.9$ and 37.91% for $y = 1.1$. Hence, the maximum percentage of Ni atoms occupying the corner positions, $P_{Ni}{}^c$, is found to be 88.21%, for $y = 0.8$, which is close to the proposed value of 80% [14, 15]. The average isomer shift values for the Fe^f and Fe^c atoms of γ'-Fe$_{4-y}$Ni$_y$N phase, change slightly from the corresponding values of γ'-Fe$_4$N i.e. 0.28 mm/s. Thus the electronic configurations of the Fe^f and Fe^c atoms, of γ'-Fe$_{4-y}$Ni$_y$N phase, are quite similar to those in Fe^f and Fe^c atoms of γ'-Fe$_4$N. The quadrupole splitting in γ'-Fe$_{4-y}$Ni$_y$N phase, have different signs for the Fe^f and Fe^c atoms, which may occur due to two different geometrical orientations. The line widths corresponding to corner sites in γ'-Fe$_{4-y}$Ni$_y$N phase, varies from 0.61 to 1.11 mm/s and this can be due to fluctuation of the exchange field due to the change of the local environment of Fe^c atoms because of the Ni atom substitutions. The hyperfine field for Fe^c atoms in γ'-Fe$_{4-y}$Ni$_y$N phase,

Springer

Table 2 ^{57}Fe Mössbauer resonance parameters for ε-Fe$_{3-x}$Ni$_x$N ($x = 0.0$–0.8)

Compositon Ni (x)	Hyperfine field ΔH_f (T)	Isomer shift δ (mm/s)	Quadrupole splitting Δ (mm/s)	Line widths Γ (mm/s)	Percent of Obs
0.0	–	0.44	0.46	0.87	–
0.1	–	0.45	0.45	0.64	–
0.2	–	0.44	0.43	0.63	–
0.3	–	0.43	0.53	0.88	–
0.4	–	0.44	0.45	0.67	–
0.5					
(sextet 1)	21.12	0.13	0.30	2.40	95.4
(sextet 2)	26.09	0.21	–	1.11	4.6
(doublet)	–	0.39	0.43	0.74	–
0.6					
(sextet 1)	21.71	0.30	0.38	1.27	90.8
(sextet 2)	24.78	0.43	−0.85	0.61	9.2
(doublet)	–	0.39	0.41	0.68	–
0.7					
(sextet 1)	22.52	0.24	0.29	1.10	87.8
(sextet 2)	24.35	0.34	−0.91	0.66	12.2
(doublet)	–	0.39	0.38	0.68	–
0.8					
(sextet 1)	22.56	0.23	0.31	1.26	79.9
(sextet 2)	24.11	0.40	−0.68	1.06	20.1
(doublet)	–	0.38	0.37	0.67	–

decrease with increasing Ni substitution form 26.09 T for $y = 0.7$ to 24.11 T for $y = 1.1$. These values are sufficiently low as compared to γ'-Fe$_4$N ($\Delta H_i = 34.5$ T) and this decrease in magnetic hyperfine field for the Fe^c atoms, is attributed to the presence of local inhomogeneities in γ'-Fe$_{4-y}$Ni$_y$N phase, due to Ni atom substitutions [14]. The hyperfine field values for the Fe^f sites in γ'-Fe$_{4-y}$Ni$_y$N phase do not change much for $y = 0.7$–1.1, and are comparable to the Fe^f value of 21.9 T for γ'-Fe$_4$N [16]. The average hyperfine field of γ'-Fe$_{4-y}$Ni$_y$N phase increase from 21.35 T for $y = 0.7$ to 22.87 T for $y = 1.1$, which is corroborated by the room temperature magnetic measurements. Due to the hyperfine field distributions, the Mössbauer spectra of γ'-Fe$_{4-y}$Ni$_y$N, is expected to split into five sextets indicating iron atoms in five different environments [14]. However, in this case since the fraction of γ'-Fe$_{4-y}$Ni$_y$N phase is very less in the ε-Fe$_{3-x}$Ni$_x$N ($x = 0.5$–0.8) compounds and hence the intensity of the sextets very small, such kind of resolution is not attained. However, similar kind of results are observed for γ'-Fe$_4$N by Saegusa et al. [17].

For ε-Fe$_{3-x}$Co$_x$N system, the magnetization vs. field (σ-H) curves do not saturate at 300 K, up to 1.1 T, indicating superparamagnetic nature while for ε-Fe$_{3-x}$Ni$_x$N, $x = 0.0$–0.4, the σ-H plots show similar superparamagnetic nature and for $x = 0.5$–0.8, the σ-H plots get saturated due to the incorporation of γ'-Fe$_{4-y}$Ni$_y$N phase. The saturation magnetization (σ_s) for all the compositions are measured from the plots of σ as a function of inverse field (H), where the intercept to the magnetization axis (from the high field strength region) as $1/H \rightarrow 0$ gives σ_s. For the Co and Ni substituted ε-Fe$_3$N system, at 300 K, the σ_s values increase from 3.65 to 170.3 emu/g for

 Springer

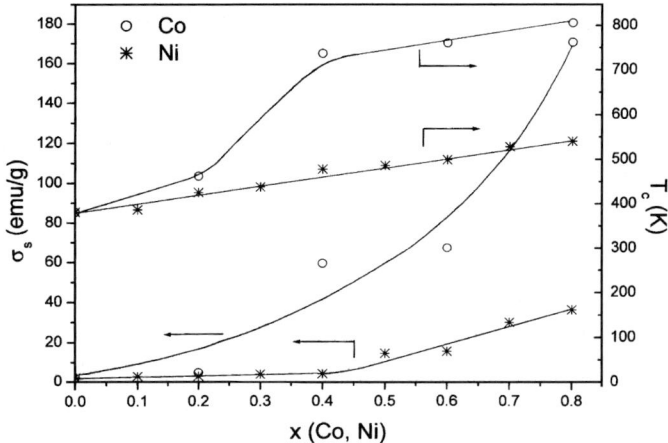

Figure 3 Variation of σ_s (300 K) and T$_c$ with x, where x is Co/Ni substituted in ε-Fe$_3$N.

$x = 0.2$–0.8 and 2.620 to 35.996 emu/g for $x = 0.1$–0.8, respectively. For ε-Fe$_{3-x}$Ni$_x$N, at 5 K, the σ_s values are almost constant from $x = 0.0$–0.4, while for $x = 0.5$–0.8, an abrupt increase in σ_s value is observed (Figure 3). This increase of σ_s values may be due to decrease of minority (spin down) d-states near Fermi energy and the increase of majority (spin up) d states. Also, the effect of the addition of Ni is to suppress the nitridation of Fe, and hence the σ_s at 300 K increases with the increase of Ni content [8]. For the representative ε-Fe$_{2.4}$Co$_{0.6}$N and ε-Fe$_{2.6}$Ni$_{0.4}$N systems, σ_s increases with the lowering of temperature from 300 to 5 K. The σ_s values for ε-Fe$_{2.4}$Co$_{0.6}$N system at 300, 70 and 5 K at an applied field of 1 T are 55.37, 86.64 and 92.93 emu/g, respectively. Similarly for ε-Fe$_{2.6}$Ni$_{0.4}$N system, the corresponding values are 3.7, 8.93 and 16.9 emu/g at 300, 70 and 5 K, respectively. The difference between the σ_s values is due to the thermal fluctuation phenomenon. With the increase of temperature, the thermal energy is increased and this results in the fluctuation of the magnetization orientation. Hence, the decrease of saturation magnetization σ_s is observed when measured at 5, 70 and 300 K. Moreover, the saturation magnetization values at 5 and 300 K are less than the reported value of 133.0 emu/g for bulk Fe$_3$N [18]. The decrease in the room temperature σ_s value may be due to the canted spin structure, spin noncollinearity and/or, oxide layer at the surface of the nanoparticles.

In ε-Fe$_{3-x}$Co$_x$N, the values of coercivity (H_c) are 0.0118, 0.0188, 0.0245, and 0.0311 T for $x = 0.0$, 0.2, 0.4, 0.6, respectively, but H_c goes down to 0.0178 T for $x = 0.8$. For ε-Fe$_{3-x}$Ni$_x$N system, at 300 K, H_c does not follow any trend with the increase of Ni content but at 5 K, H_c increases from 0.0353 T for $x = 0.1$ to 0.0434 T for $x = 0.4$ and again decreases to 0.0234 T for $x = 0.8$. This type of behavior is reported for γ'-Fe$_{4-x}$Ni$_x$N system where the maximum coercive field is observed for $x = 0.4$ [19]. H_c depends mainly on two factors: a) Geometrical nature related to size and shape of particles and b) magnetocrystalline anisotropy [20]. They are also found to depend on the distribution of particle size. Since these systems belong to the hcp crystal structure, geometrical nature is believed to have the major effect in determining the value of H_c. Because of different particle size distribution in these systems, a definite fraction of the nanocrystalline particles lie in the critical size range corresponding to single domain. This results in larger values of coercivity because

 Springer

of the reversal of magnetization by spin rotation mechanism only, which has been calculated by Stoner and Wohlfarth [21–23]. Decrease in coercivity can be due to the presence of particle size ranges either larger or smaller than the critical single domain size range.

The Curie temperature (T_c) is found to increase with the increase of Co and Ni in ε-Fe$_3$N system. For ε-Fe$_{3-x}$Co$_x$N and ε-Fe$_{3-x}$Ni$_x$N systems, T_c increases from 463 K ($x = 0.2$) to 806 K ($x = 0.8$) and from 387 K ($x = 0.1$) to 540 K ($x = 0.8$), respectively (Figure 3). The increase of T_c in Co substituted system is more than the Ni substituted system and this implies more magnetic interactions in the former. The increase in T_c may be attributed to large exchange energy interactions between Fe, (Co, Ni) and N atoms and the decrease in the number of negative Fe–Fe exchange bonds as a result of increasing interatomic distance, which in turn increase the magnitude of ferromagnetic exchange between Fe–Fe atoms [12, 20, 24].

The zero-field cooled (ZFC) and field-cooled (FC) magnetization curves are measured for the ε-Fe$_{3-x}$Ni$_x$N, $x = 0.0$–0.8, system between the temperature range of 10–300 K, at an external applied field of 0.025 T. Blocking of these nanoparticles occur near room temperature, at the blocking temperature (T_B), where the ZFC and FC magnetization curves bifurcate from each other. Above T_B, the particle moments are in thermal equilibrium and show superparamagnetic nature and below T_B, the moments are ferromagnetically blocked. For ε-Fe$_{3-x}$Ni$_x$N, $x = 0.5$–0.8, with the increase of temperature, the ZFC magnetization value continuously increases till 300 K, while the FC curve steadily decreases and finally merge with ZFC near 300 K. The characteristic behavior of the ZFC and FC curves for the compositions, $x = 0.5$–0.8, clearly indicates ferromagnetic nature of the particles at room temperature, which arises due to the incorporation of ferromagnetic γ'-Fe$_{4-y}$Ni$_y$N phase into the superparamagnetic ε-Fe$_{3-x}$Ni$_x$N phase. Such kind of behavior of ZFC and FC magnetization curves has already been reported in the literature for MnFe$_2$O$_4$ sample [25].

he ZFC and FC magnetization curves are measured at various external applied fields within the temperature range 10–300 K, for the representative systems ε-Fe$_{2.4}$Co$_{0.6}$N and ε-Fe$_{2.6}$Ni$_{0.4}$N systems as shown in Figure 4. The corresponding T_B values are found to decrease as the applied field is increased. This is due to the fact that as the applied field increases, the anisotropy energy of the particles decreases and hence the particles need less thermal energy to cross over the energy barrier. The irreversibility observed in these curves is attributed to the interparticle interactions and this irreversibility gets reduced with the increase of external applied field and finally almost vanishes at 1.0 T for ε-Fe$_{2.6}$Ni$_{0.4}$N nanoparticles. For ε-Fe$_{2.4}$Co$_{0.6}$N and ε-Fe$_{2.6}$Ni$_{0.4}$N nanoparticles, the ZFC and FC curves display peaks in the range 15–70 K, after which it again starts to decrease. Due to the unidirectional pinning of the FM spins by the AF layer at the surface, different domains are created with individual exchange anisotropy. When the sample is cooled down to a sufficiently low temperature, the domain states are frozen in with random exchange anisotropy and because of this frozen state, the domain states get stabilized. This pinning effect is caused by the creation of a local unidirectional exchange anisotropy that follows the domain magnetization directions and this combined effect gives the peak at low temperature in the ZFC curves.

The low temperature plateau or maximum in the FC curve for these samples is similar to that observed in a spin-glass system and hence these systems are expected to display spin-glass like nature. The origin of the spin-glass like nature is believed to

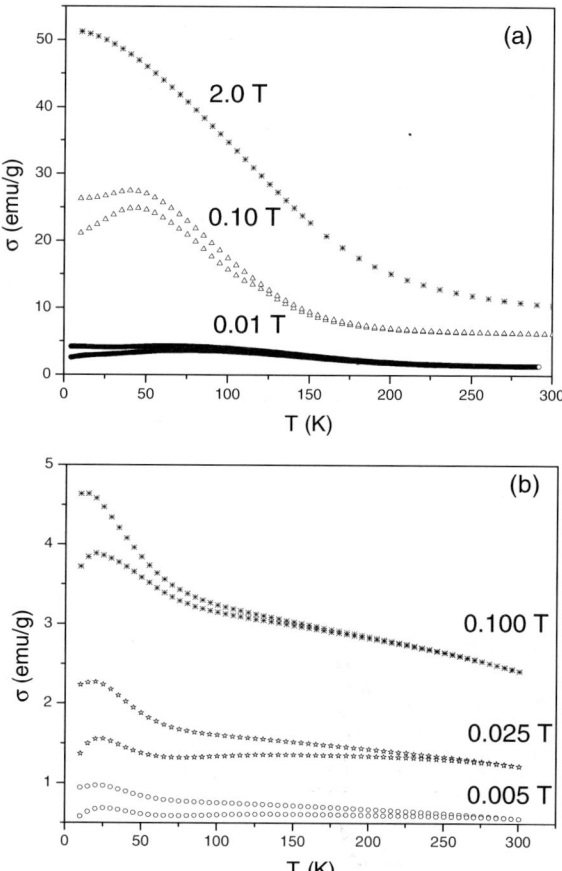

Figure 4 ZFC and FC plots at various external applied fields within the temperature range 10–300 K, for the representative systems **a** ε-Fe$_{2.4}$Co$_{0.6}$N and **b** ε-Fe$_{2.6}$Ni$_{0.4}$N.

be due to the freezing of the spins at the surface of the nanoparticles. The interface between the antiferromagnetic (AF) oxide layer and the ferromagnetic (FM) nitride core is not clearly defined and may have multiple random interfaces. Hence different domains with AF–FM interactions are created and surface oxidation occurs not only at the grains but also through the grain boundaries [26]. Hence a disordered state is created with the occurrence of continuous spin-flop-like process when each domain interacts with neighboring domains of oppositely oriented anisotropy fields. However, the exchange type interactions between the surface atoms of neighboring particles may be an additional effect to the disordered state. These incidents account for the demagnetizing effects and spin canting at the surface of the nitride nanoparticles which results into freezing of surface spins at low temperature. However, the role of some other effects influencing the spin-glass like ordering cannot be ruled out.

4. Conclusions

Co and Ni substituted ε-Fe$_3$N nanoparticles are synthesized using precursor routes with nitridation in NH$_3$ atmosphere. X-ray diffraction patterns are indexed with

Rietveld analysis based on the hexagonal ε-Fe$_3$N phase with the space group P6$_3$/mmc. For ε-Fe$_{3-x}$Co$_x$N system, $x = 0.4$–0.8, a mixture of hexagonal ε-Fe$_3$N and bcc α-Fe phases are formed and for ε-Fe$_{3-x}$Ni$_x$N, $x = 0.5$–0.8, small amount of cubic γ'-Fe$_{4-x}$Ni$_x$N is also precipitated. Co and Ni randomly substitute in Fe sites of ε-Fe$_3$N phase and the XRD crystallite size is found to be in the range 10–20 nm. The ^{57}Fe Mössbauer studies and magnetic studies indicate the superparamagnetic nature of these particles. ZFC/FC curves show the strong magnetic interactions among these particles. The intrinsic magnetic properties are found to change with the change in Co and Ni contents. The low temperature magnetic properties are interpreted in the light of spin-glass like ordering and exchange bias phenomena at the surface of these nitride nanoparticles.

Acknowledgments   The authors wish to acknowledge MHRD, New Delhi for the financial support.

References

1. Toth, L.E.: In: Transition Metal Carbides and Nitrides, New York (1971)
2. Oyama, S.T. (ed.): In: The Chemistry of Metal Carbides and Nitrides. Blackie Academic, London (1996)
3. Matar, S., Siberchicot, B., Penicaud, M., Demazeau, G.: J. Phys. I France **2**, 1819 (1992)
4. Jack, K.H.: Proc. Roy. Soc. A **208**, 200 (1951)
5. Siberchicot, B., Matar, S.F., Fournes, L., Demazeau, G., Hagenmuller, P.: J. Solid State Chem. **84**, 10 (1990)
6. Andriamandroso, D., Fefilatiev, L., Demazeau, G., Fournes, L., Pouchard, M.: Mat. Res. Bull. **19**, 1187 (1984)
7. Chen, S.K., Jin, S., Tiefel, T.H., Hsieh, Y.F., Gyordy, E.M., Johnson, Jr., D.W.: J. Appl. Phys. **70**, 6247 (1991)
8. Takahashi, S., Umeda, K., Kita, E., Tasaki, A.: IEEE Trans. Magn. **MAG23**, 3630 (1987)
9. Takahashi, S., Kume, M., Matsuura, K.: IEEE Trans. Magn. **26**, 1632 (1990)
10. Bozorth, R.M.: Ferromagnetism. Van Nostrand, New York (1951)
11. Gajbhiye, N.S., Bhattacharyya, Sayan: Phys. Stat. Sol. (c) **1**, 3764 (2004)
12. Li, H., Coey, J.M.D.: In: Buschow, K.H.J. (ed.) Handbook of Magnetic Materials, vol. 6. North-Holland, Tokyo (1991)
13. Gajbhiye, N.S., Ningthoujam, R.S., Weissmuller, J.: Hyper. Interact. **156**, 51 (2004)
14. Panda, R.N., Gajbhiye, N.S.: J. Magn. Magn. Mater. **195**, 396 (1999)
15. Shirane, G., Takei, W.J., Ruby, S.L.: Phys. Rev. **126**, 49 (1962)
16. Panda, R.N., Gajbhiye, N.S.: IEEE Trans. Mag. **34**, 542 (1998)
17. Saegusa, N., Morrish, A.H., Tasaki, A., Tagawa, K., Kita, E.: J. Magn. Magn. Mat. **35**, 123 (1983)
18. Robbins, M., White, J.G.: J. Phys. Chem. Solids **25**, 717 (1964)
19. Jonsson, T., Svedlindh, P., Hansen, M.F.: Phys. Rev. Lett. **81**, 3976 (1998)
20. Mulder, C.A.M., van Duyneveldt, A.J., Mydosh, J.A.: Phys. Rev. B **23**, 1384 (1981)
21. Martinez, B., Obradors, X., Balcells, Ll., Rouanet, A., Monty, C.: Phys. Rev. Lett. **80**, 181 (1998)
22. Jonsson, T., Jonason, K., Jonsson, P., Nordblad, P.: Phys. Rev. B **59**, 8770 (1999)
23. Luis, F., del Barco, E., Hernandez, J.M., Remiro, E., Bartolome, J., Tejada, J.: Phys. Rev. B **59**, 11837 (1999)
24. Mydosh, J.A.: Spin Glass: An Experimental Introduction. Taylor & Francis, London (1993)
25. Balaji, G., Gajbhiye, N.S., Wilde, G., Weissmuller, J.: J. Magn. Magn. Mater. **242–245**, 617 (2002)
26. Nogués, J., Ivan, K., Schuller, J.: J. Magn. Magn. Mater. **192**, 203 (1999)

Hyperfine Interact (2005) 164: 27–33
DOI 10.1007/s10751-006-9230-8

La–Zn Substituted Hexaferrites Prepared by Chemical Method

A. Grusková · J. Lipka · M. Papánová · J. Sláma ·
I. Tóth · D. Kevická · G. Mendoza · J. C. Corral ·
J. Šubrt

Abstract La–Zn substituted M-type Ba hexaferrite powders were prepared by sol-gel (Mx) and organometallic precursor (Sk) methods with Fe/Ba ratio of 11.6 and 10.8, respectively. The compositions $(LaZn)_x Ba_{1-x} Fe_{12-x} O_{19}$ with $0.0 \leq x \leq 0.6$ were annealed at 975°C/2 h. The cationic site preferences of nonmagnetic La^{3+} instead of Ba^{2+} ions and Zn^{2+} instead of Fe^{3+} ions were determined by Mössbauer spectroscopy. The La^{3+} ions substitute the large Ba^{2+} ions at 2a site and for $x \geq 0.4$ also at $4f_2$ site. The nearly all Zn^{2+} ions are placed at the $4f_1$ sites. The thermomagnetic analysis of $\chi(\vartheta)$ confirms that only the small substitutions for $x \leq 0.4$ can be taken as a single-phase hexaferrites. The coercivity H_c almost does not change at $x = 0.2$ for (Mx) samples and further decrease up to $x = 0.6$. For (Sk) samples at substitution $x = 0.2$ the values of H_c are decreasing and at higher x the values nearly do not change. The Curie points, T_c, slowly decrease with x for both (Mx) and (Sk) samples.

Key words ferrites-hexagonal · magnetic recording · Mössbauer effect.

1. Introduction

The substituted Ba hexaferrites are potential candidates for recording media because of their mechanical hardness and chemical stability. In order to improve the

A. Grusková · J. Lipka (✉) · M. Papánová · J. Sláma · I. Tóth · D. Kevická
Slovak University of Technology, Ilkovičova 3, 812 19 Bratislava, Slovakia
e-mail: jozef.lipka@stuba.sk, lipkajozef@cdicon.sk

A. Grusková
e-mail: anna.gruskova@stuba.sk

G. Mendoza · J. C. Corral
Cinvestav-Saltillo, P.O. Box 663, 25900 Saltillo, Coah, Mexico

J. Šubrt
Institute of Inorganic Chemistry, AS CR, 250 68 Řež, Czech Republic

fundamental magnetic properties of hexaferrite, many studies have been carried out on synthesis methods and cationic substitutions of divalent or multivalent ions and of their mixture. In M-type hexaferrite the iron ions are positioned on five non-equivalent sites 2a, 12k and $4f_2$ octahedral, $4f_1$ tetrahedral and 2b bipyramidal sites. In the magnetically ordered state in Ba ferrite the 12k, 2a and 2b sites have their spins aligned parallel to each other in the crystallographic c-axis, whereas those of $4f_2$ and $4f_1$, point into the opposite direction. The magnetic properties of the substituted hexaferrites strongly depend on the electronic configuration of the substituting cations. It is known, that more electronegative ions prefer octahedral coordination [1]. The electronegativity of substituted ions La^{3+}, Zn^{2+} is 1.10 and 1.65, respectively. According to the ligand field [2], ions with d^1, d^2, d^3 and d^4 orbitals prefer tetrahedral and ions with d^6, d^7, d^8 and d^9 orbitals occupy octahedral positions mainly. Ions with d^0, d^5, d^{10} orbitals have no site preference. The tendency to occupy a particular site depends also on the ionic radii of the ions and their partner cations.

The investigations have been performed covering two aspects of preparations, which are Fe/Ba ratio and heat treatment [3–5]. At the sol-gel preparation of Ba ferrite, when the Ba/Fe is higher than the 11.5, nonmagnetic $BaFe_2O_4$ is presented [6]. In the previous work [7] by the precursor method a Fe/Ba ratio of 10.8 were used. Different compounds as starting materials have different solubilities in different aqueous media, so that the range of Fe/Ba values found yielding single-phase materials, could be related to these aspects. The optimum Fe/Ba ratio depends on the raw materials used and the processing procedure [6]. The substitution Ba^{2+} by La^{3+} ions is associated with a valence change of Fe^{3+} to Fe^{2+} at 2a or $4f_2$ site [8]. A suitable content of La–Zn substitution can reduce the value of $\Delta H_c/\Delta T$ near room temperature [9]. La–Zn substitutions show clearly their positive effect on J_{s-m}. The coercivity H_c decreases much slower, it changes from 464 to 224 kA/m, where $0 \leq x \leq 0.6$ [3].

The presented investigation reports on the magnetic properties and determination of the crystallographic site of La–Zn substitutions. For the preparation of the samples (Mx) by sol-gel method with traditional evaporation drying procedure and (Sk) by modified organometallic precursor method with alcohol drying were employed.

2. Experimental

Samples of $(LaZn)_xBa_{1-x}Fe_{12-x}O_{19}$ (Mx) with $0.0 \leq x \leq 0.6$ were prepared by sol-gel way [3]. $Fe(NO_3)_3.9H_2O$, $BaCO_3$, $ZnCl_2$ and $La(NO_3)_3.6H_2O$ and citric acid were used as starting materials with a high purity of 99.99%. A Fe/Ba ratio of 11.6 was used. The obtained powders were annealed at 975°C for 2 h.

Samples of $(LaZn)_xBa_{1-x}Fe_{12-x}O_{19}$ (Sk) with $0.0 \leq x \leq 0.6$ were prepared by organometallic precursor way [5]. $Fe(NO_3)_3.9H_2O$, $Ba(OH)_2.8H_2O$, $La(NO_3)_3.6H_2O$, further ZnO dissolved in HNO_3, and citric acid were used as the starting materials, all of 99% purity. In this case a Fe/Ba ratio of 10.8 was used. The samples were annealed at temperatures of 700 and 975°C for 2 h in a muffle furnace.

The magnetic properties were studied according to [10], using a vibrating sample magnetometer with a maximum external magnetic field up to 540 kA/m. The phase constitution was analyzed by Mössbauer spectroscopy using a conventional constant acceleration equipment with source of ^{57}Co in Rh matrix. The temperature dependencies of the magnetic susceptibility $\chi(\vartheta)$ were measured by the bridge method

 Springer

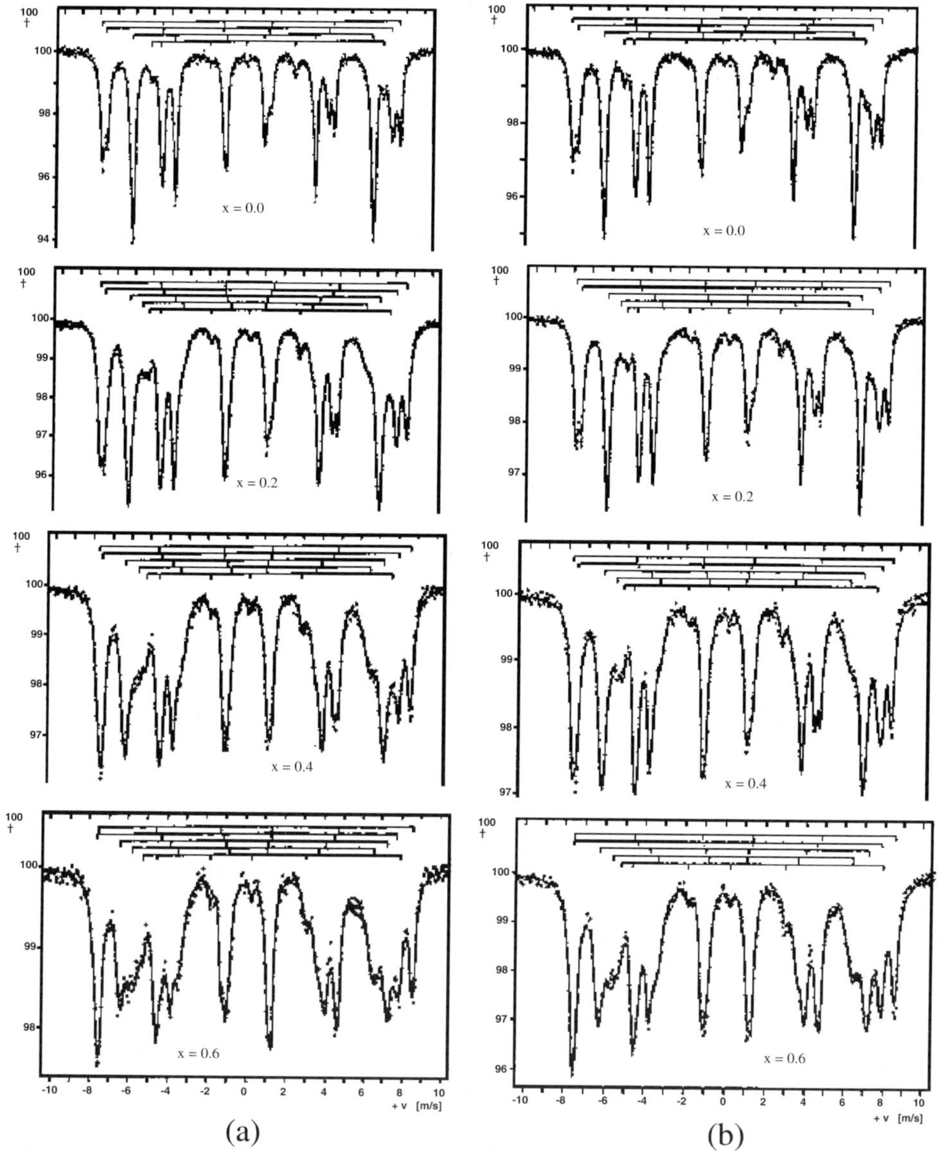

Figure 1 The room temperature Mössbauer spectra for La–Zn substituted BaM, for **a**: (Mx) and **b**: (Sk) samples, for $0.0 \leq x \leq 0.6$.

according to [11]. A Philips XL 30 scanning electron microscope (SEM) was used to obtain data for microstructural characteristics.

3. Results and discussion

The room temperature Mössbauer spectra of La–Zn substituted BaM samples prepared by sol-gel method (Mx) with $x = 0.0, 0.2, 0.4$ and 0.6 are shown in Figure 1a

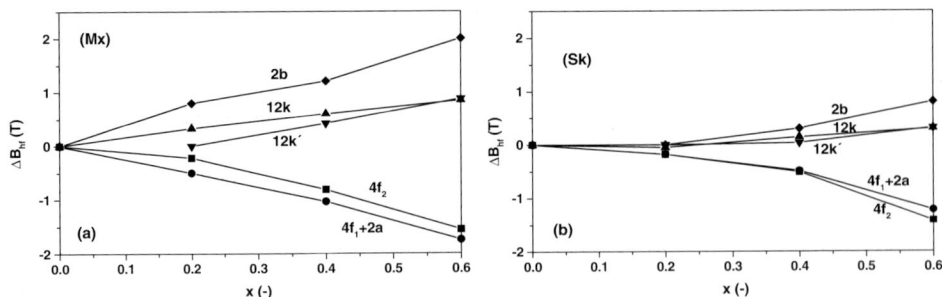

Figure 2 ΔB_{hf} vs. x, for La–Zn substituted BaM, for **a**: (Mx) and **b**: (Sk) samples.

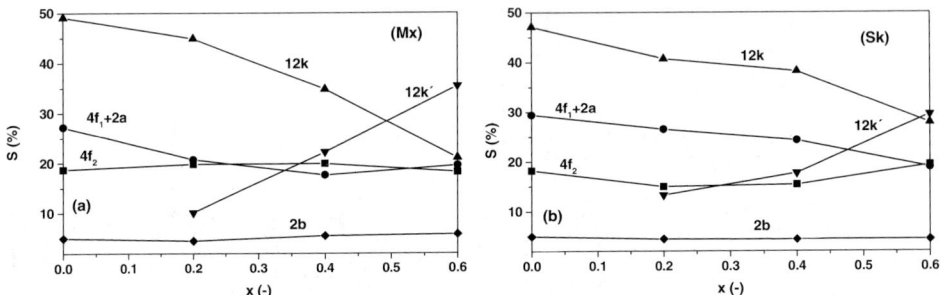

Figure 3 The relative areas S (%) vs. x, for La–Zn substituted BaM, for **a**: (Mx) and **b**: (Sk) samples.

and same composed samples prepared by precursor method (Sk) are in Figure 1b. The spectra were fitted with four sextets corresponding to $4f_2$, $4f_1 + 2a$, 12k and 2b, for $x = 0.0$. The hyperfine parameters for 2a and $4f_1$ sites are nearly equal, so that subpatterns could not be resolved. The substitutions of La–Zn ions from 0.2 up to 0.6 on the Fe^{3+} sites cause the appearance of five different magnetic surroundings and the 12k position is broadened. This position splits into two sublattices, one representing the ions with three nearest magnetic neighbours in the $4f_1$ site and other, 12k′ with two neighbours in the $4f_1$ site whereby its value B_{hf} is the smallest at about 36 T [5, 12]. The variations of local hyperfine fields ΔB_{hf} (T) and of the relative areas S (%) as a function of x for both (Mx) and (Sk) samples are shown in Figures 2 and 3. It can be seen clearly that diamagnetic La^{3+} and Zn^{2+} ions show remarkable preferences for the sites $4f_1 + 2a$, $4f_2$ and also 12k for sample (Mx), as it can be seen in Figure 2a. The same site preferences are also shown in Figure 2b for the samples (Sk), but slightly more are occupied the $4f_2$ rather than $4f_1 + 2a$ sites.

There is agreement with [8], where La^{3+} is associated with a valence change of Fe^{3+} to Fe^{2+} at 2a or $4f_2$ site and nearly all Zn ions are placed at the tetrahedral $4f_1$ site. This position has a negative contribution to magnetic polarisation because substitution of Fe^{3+} ion with spin-down by non-magnetic Zn^{2+} ions, increase resultant values of J_{s-m}.

Sauer [13] found, that in La–Zn the expected Fe^{2+} should reside on the 2a sites. At higher concentrations ($x \geq 0.6$) we detect strong reduction of the quadrupole splitting QS (mm/s) in the $4f_1 + 2a$ sites, and the relative area S (%) for 12k′ position

Table I Magnetic properties of La–Zn substituted BaM

$x(-)$		J_{s-m} $(10^{-6}\,\mathrm{T\,m^3\,kg^{-1}})$	J_{s-r} $(10^{-6}\,\mathrm{T\,m^3\,kg^{-1}})$	H_c (kA/m)	$\Delta H_c/\Delta T$ (kA/m°C)
0.0	Sol-gel	66.11	28.59	400	0.40
0.2		73.14	34.95	395	0.35
0.4		80.83	41.41	370	0.35
0.6		81.16	41.58	325	0.26
0.0	Precursor	68.85	37.07	375	0.60
0.2		93.83	48.07	330	0.46
0.4		76.27	39.19	340	0.32
0.6		74.85	38.98	340	0.40

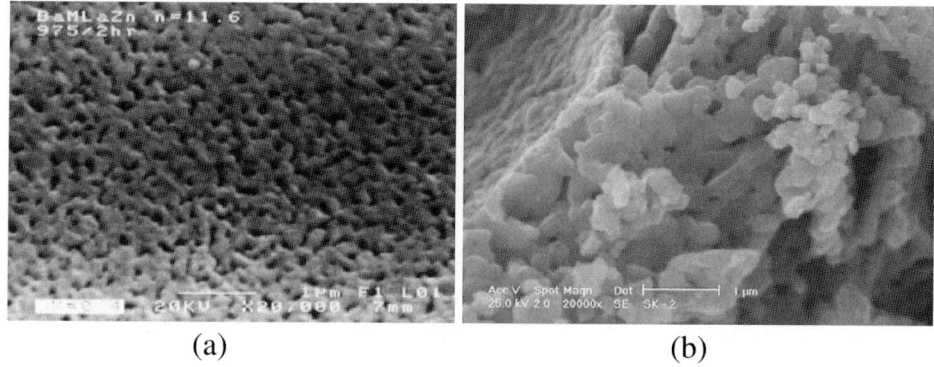

(a)   (b)

Figure 4 The micrograph of the La–Zn substituted BaM, for $x = 0.4$, for **a**: (Mx) and **b**: (Sk) samples.

increases, whilst that decreases for 12k for both (Mx) and (Sk) samples, as is shown in Figure 3a, b.

The magnetic properties of La–Zn (Mx) and (Sk) samples with $0.0 \leq x \leq 0.6$, measured at room temperature are summarized in Table I. The results of the specific magnetic polarisation J_{s-m} and J_{s-r} show the positive effect of the La–Zn substitution on these values, with concentration rate in (Mx) samples. The coercivity H_c decreases mid-slowly, it varies from 400 to 325 kA/m, where x changes from 0.0 to 0.6. The H_a values are strongly dependent on contents of the La^{3+} ions [8]. For (Sk) samples J_{s-m}, J_{s-r} expressively increase for $x = 0.2$; further at higher substitution the values are decreasing. The coercivity H_c at $x = 0.2$ decreases, at $x = 0.4$ slightly increases or it does not change. It can be seen, that the best values of J_{s-m}, J_{s-r} and of H_c are achieved at $x = 0.2$ for (Sk) samples. The values of temperature coefficient of coercivity $\Delta H_c/\Delta T$, defined in [8] are shown in Table I. The La–Zn (Sk) samples have positive values of $\Delta H_c/\Delta T$, which are slightly higher than those of the La–Zn (Mx) samples.

Micrographs taken by SEM of La–Zn (Mx) and (Sk) samples with $x = 0.4$ are in Figure 4a, b. It was found [3] that, La–Zn substitution leads to 'D' values clearly of about 40 nm and 't' decreases to about 25 nm up to $x = 0.4$, the aspect ratio (D/t) varies between 2.3 and 3.2. The La–Zn substitution yields a homogeneous microstructure and finer crystalline size in (Mx) samples, than those in (Sk) samples. Cell

 Springer

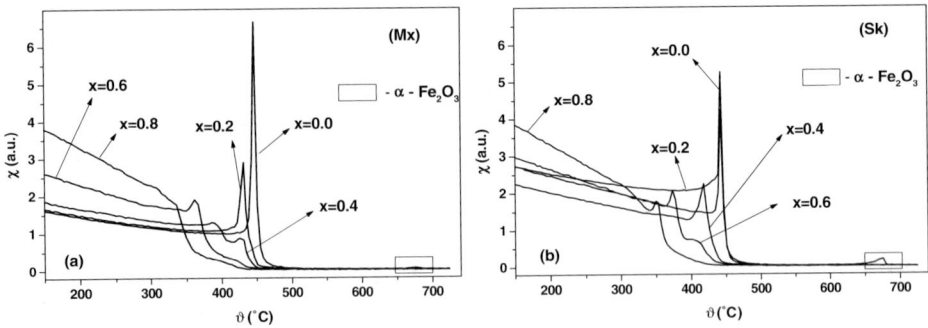

Figure 5 The $\chi(\vartheta)$ dependencies for La–Zn substituted BaM, for **a**: (Mx) and for **b**: (Sk) samples.

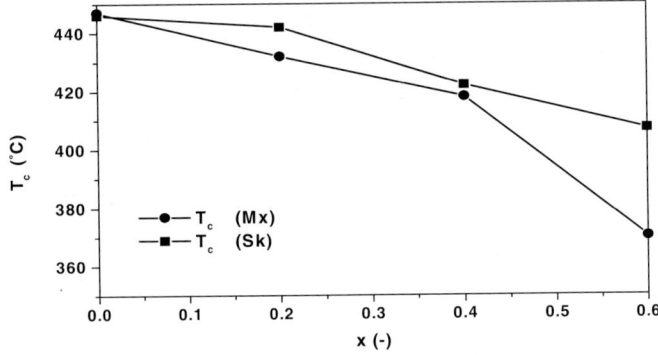

Figure 6 The Curie temperature dependencies of concentration rate for La–Zn substituted BaM samples.

parameters examination show that c-axis decreased markedly and a-axis decrease for $x < 0.4$, whilst an increase are observed for $x > 0.4$. According to [9] a- and c-axes decreased with x. These discrepancies might be related to way of the processing.

When the substitutions are of $x \leq 0.4$, the samples form single phase M-type hexagonal ferrite, what was confirmed by the measurements of the magnetic susceptibility. The $\chi(\vartheta)$ dependencies for La–Zn substituted (Mx) and (Sk) samples are shown in Figure 5a, b.

It can be seen that at room temperature, χ increases with the substitution level. In the vicinity of the Curie temperature T_c, a sharp Hopkinson peak occurs for $x = 0.0$ and for $x = 0.2$ at (Mx) samples. At $x \geq 0.6$ there are multiphase systems. The (Sk) samples have Hopkinson peak continuously up to $x = 0.4$. There is presently single-phase system with homogeneous microstructure. The samples with $x \geq 0.6$ were detected as unstable multiphase systems [14]. Here was occurrence of the hematite phase that was confirmed by Mössbauer spectroscopy measurements. For example, for $x = 0.8$, the composition contained also secondary hematite phase, (α-Fe$_2$O$_3$) with the relative area of 12.2% for (Mx) and of 33.8% for (Sk) samples further the hexaferrite phase. The T_c decreases linearly with increasing La–Zn substitutions, by only about of 15% for (Mx) samples and of 7% for (Sk) samples, Figure 6.

 Springer

4. Conclusions

The magnetic analysis of La–Zn substituted M-type Ba ferrites prepared by the sol-gel method showed better J_{s-m}, J_{s-r} and H_c than those prepared by the precursor method. The J_{s-m}, J_{s-r} increase for (Mx) samples up to $x = 0.4$ and then slowly at higher substitutions. The La–Zn substituted (Sk) samples have J_{s-m}, J_{s-r} values increasing only up to $x = 0.2$, then their values decreases at $x \geq 0.4$ and further are almost not changing. By La–Zn ions is probably more substituted 2a site than those of $4f_1$ and of $4f_2$ sites. We found, that Zn^{2+} occupies the $4f_1$ sites, which have negative contribution to magnetic polarisation, what might be confirmed by slight increase of J_{s-m} and J_{s-r} at small x. The H_c as a function of x, slowly decreases at both substitutions. The expressive decrease of H_c at $x = 0.2$ for (Sk) sample confirmed, that substitutions of La^{3+} ions for Ba^{2+} ions is associated with a valence change of Fe^{3+} to Fe^{2+} ion at 2a or $4f_2$ site.

Acknowledgment We thank VEGA of the Slovak Republic under project No. G-1/3096/06 and CONACYT – Mexico under project J28283U.

References

1. Rane, M.V., et al.: J. Magn. Magn. Mater. **195**, L256 (1999)
2. Rane, M.V., et al.: J. Magn. Magn. Mater. **192**, 288 (1999)
3. Corral, J.C., et al.: J. Magn. Magn. Mater. **242–245**, 430 (2002)
4. Zhong, W., et al.: J. Magn. Magn. Mater. **168**, 196 (1997)
5. Grusková, A., et al.: J. Magn. Magn. Mater. **242–245**, 423 (2002)
6. Mendoza-Suárez, G., et al.: Mater. Chem. Phys. **9454**, 1 (2002)
7. Lipka, J., et al.: J. Magn. Magn. Mater. **140–144**, 2209 (1995)
8. Liu, X., et al.: J. Magn. Magn. Mater. **238**, 207 (2002)
9. Liu, X., et al.: J. Appl. Phys. **87**(5), 2503 (2000)
10. Dosoudil, R.: J. Electr. Eng. **53**(10/S), 135 (2002)
11. Jančárik, V., et al.: J. Electr. Eng. **50**(8/S), 63 (1999)
12. Zhou, J.X., et al.: IEEE Trans. Magn. **27**(6), 4654 (1991)
13. Sauer, Ch., et al.: J. Phys. Chem. Solids **39**, 1197 (1978)
14. Turilli, G., et al.: IEEE Trans. Magn. **24**, 2865 (1988)

Hyperfine Interact (2005) 164: 35–40
DOI 10.1007/s10751-006-9232-6

Preparation and Properties of Iron and Iron Oxide Nanocrystals in MgO Matrix

O. Schneeweiss · R. Zboril · N. Pizurova · M. Mashlan

Abstract We have prepared α-iron and magnetite (Fe_3O_4) nanoparticles in MgO matrix from a mixture of nanocrystalline Fe_2O_3 with Mg(H,O) powders calcinated in hydrogen. This procedure yielded spherical magnetic nanoparticles embedded in MgO. Transmission electron microscopy and Mössbauer spectroscopy were used for structure and phase analysis. The measurements of magnetic properties showed increased coercivity of the nanocomposite samples.

Keywords nanocrystalline materials · magnetite · α-Fe · nanocomposite · magnetic properties.

1. Introduction

Nanocrystalline composites are target of an intensive technological research which is focused on the methods of preparation of these materials in large scale and for a reasonable low price. The mechanical alloying or mechanosythesis give the materials in the largest amount but there are some limits in particle size which are formed as conglomerates of large number of tight bonded nanocrystals of a single phase. Therefore new technologies are target of the research in this field where separated nanocrystals can be obtained [1, 2].

The nanocomposites for applications in magnetic circuits are composed of magnetic (ferro- or ferrimagnetic) nanocrystals and a matrix (electrical insulator) which separates the magnetic particles (grains). The main nanocomposite processing steps usually consist of synthesis of magnetic and matrix nanoparticles, their mixing, and

O. Schneeweiss (✉) · N. Pizurova
Institute of Physics of Materials, AS CR, Žižkova 22, 61662 Brno, Czech Republic
e-mail: schneew@ipm.cz

R. Zboril · M. Mashlan
Faculty of Science, Palacký University, tř. Svobody 26, 770 46 Olomouc, Czech Republic

Table I Magnetic properties of pure iron and magnetite Fe_3O_4 [5, 6]

	Fe	Fe_3O_4
Saturation magnetization [A m^2 kg^{-1}]	218	$92 \div 100$
Anisotropy constant [Jm^{-3}]	105	$104 \div 105$
Magnetostriction constant [10^6]	25	35
Curie temperature [K]	1,044	850

consolidation or compaction of the insulated nanoparticles (powder) into a solid of near theoretical density to produce bulk materials with different shapes and sizes. The matrix electrically passivates the particle and serves following important purposes: (a) Prevents oxidation since the particles are extremely chemically reactive and pyrophoric at ambient conditions, (b) develops a significant barrier to eddy currents by increasing the nanostructure resistivity, and (c) hinders grain growth or particle agglomeration of the insulated particles during compaction and subsequent exploitation at high temperature.

There is an important requirement for a nanocomposite magnetic material which must exhibit good soft magnetic properties – the magnetic moments of neighbouring particles be magnetically coupled by what is known as magnetic moment exchange coupling [3, 4]. A critical parameter, the exchange coupling length, is the distance within which the magnetic moments of the neighbouring particles can be coupled. If the distance between the neighbouring particles is greater than the exchange length, then a nanocomposite magnetic solid will result with poor soft magnetic properties. This can be caused by several factors during the preparation of the material, e.g. large particle size of the matrix material, poor mixing of the components and/or poor compaction of the mixed powder. Therefore all these factors must be capable of making the particle separation distance less than the critical exchange coupling length. It should be noted, that each magnetic alloy has its own critical exchange coupling length. To make a design of components of magnetic circuits specific knowledge about the electrical, magnetic, and thermal properties of the soft magnetic materials used in these components is required.

In this work we have chosen as the magnetic particles pure iron and magnetite. They are known as excellent magnetic properties and their physical properties are well described (Table I). They also have good biocompatibility. The nanoparticles of these materials have however some disadvantage. The iron nanocrystals oxides very rapidly even at room temperature and therefore the matrix must protect them against oxidation or another reaction with the surrounding medium up to higher temperatures of their utilization. Besides that the matrix must be stable and to effect against grain coarsening (growth) and aggregation.

For the matrix we have tested MgO. It has a low specific weight, high chemical and temperature stability and good lattice compatibility with the nanocrystals of the magnetic particles.

2. Experimental details

The nanocomposites were prepared from the amorphous (nanocrystalline $\sim$3 nm) Fe_2O_3 prepared by chemical processing [7] and nanocrystalline Mg(O,H) powder

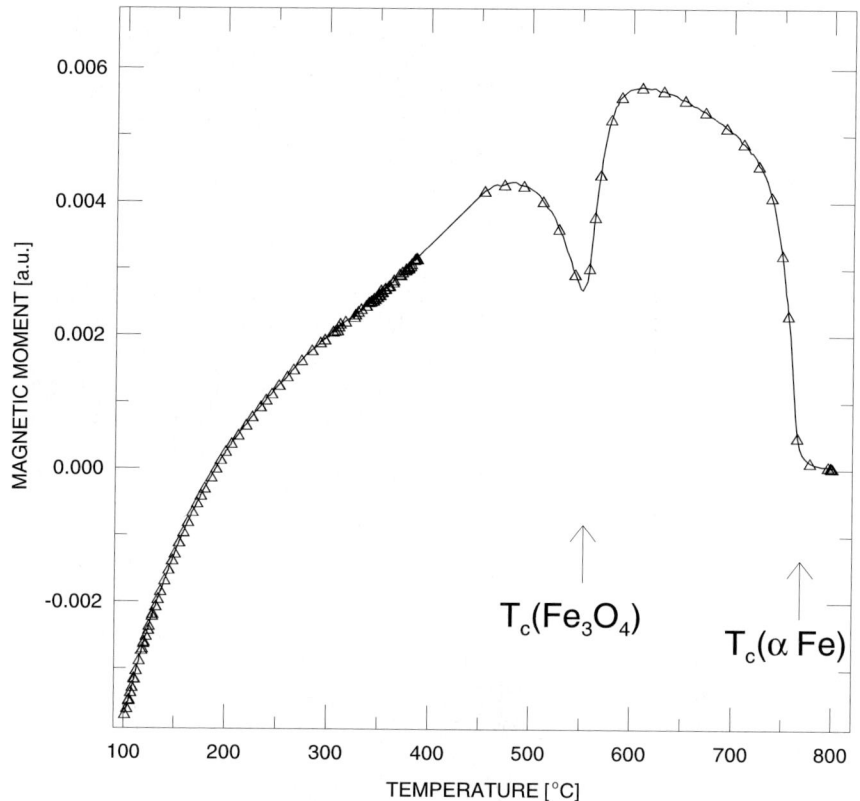

Figure 1 Temperature dependence of the magnetic moment of the as-mixed powder Fe_2O_3 with Mg(O,H).

prepared by spark erosion technique [8]. The components were carefully mixed in a ball mill equipment (quartz balls in a polyethylene vial), compacted by pressing at room temperature and annealed subsequently in vacuum and hydrogen atmosphere. The changes in phase composition were studied using temperature dependence of magnetic moment. For these measurements vibrating sample magnetometer and external field 5 mT was used. The phase composition was investigated using X-ray diffraction, ^{57}Fe Mössbauer spectroscopy, and transmission electron microscopy (TEM).

3. Results and discussion

The thermomagnetic curves for are shown in Figure 1. Two main stages of transformation of the as mixed material are indicated. The first transformation occurs below Curie temperature of magnetite with a beginning at ~570 K. Therefore the nanocomposite consisting of Fe_3O_4 in MgO was prepared by heating of compacted pellets of the original powders at 570 K for 1 h. The low temperature of the heat treatment was chosen for holding as low as possible grain growth. The TEM pictures

Figure 2 TEM of the Fe_3O_4 in MgO sample.

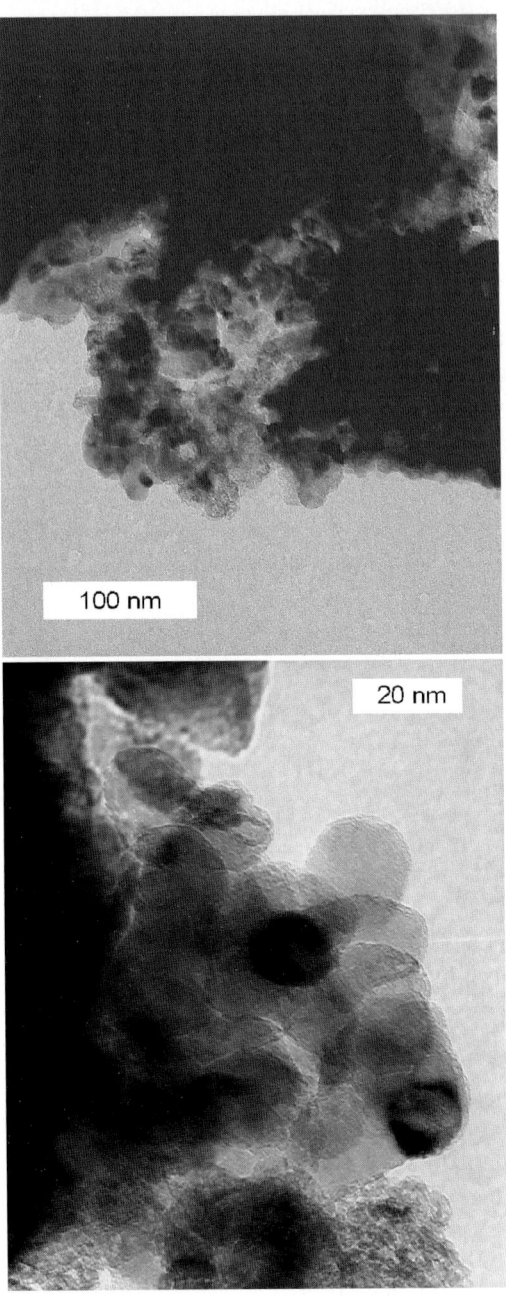

(Figure 2) show that the composite consists of 20 ± 5 nm spherical magnetite nanoparticles and MgO debris. The phase analysis based on Mössbauer spectra (Figure 3) measured at room temperature shows clearly magnetite patterns with parameters corresponding to the bulk magnetite. The shape of the hysteresis loop (Figure 4) corresponds well to the isolated magnetic particles. The coercive force

 Springer

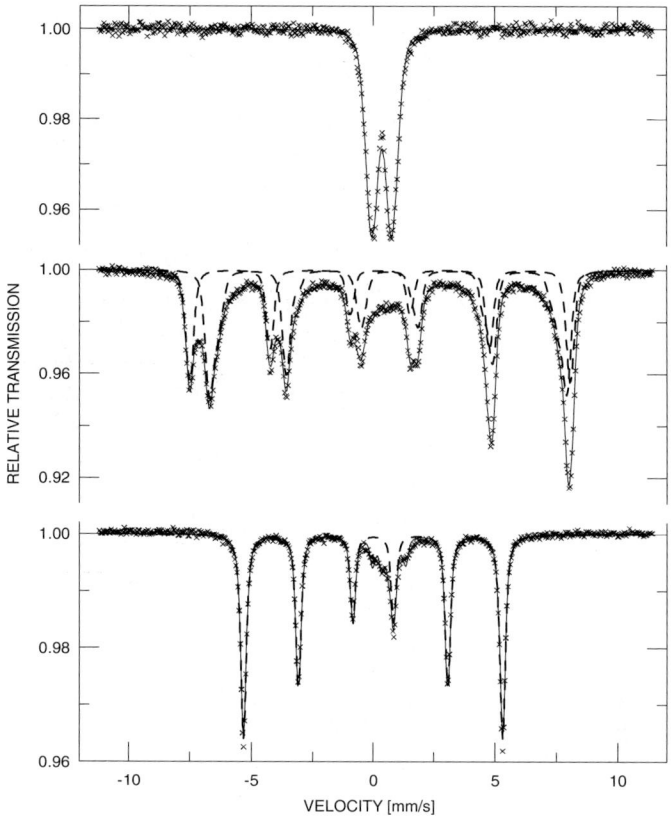

Figure 3 Mössbauer spectra of the as-mixed powder Fe_2O_3 with $Mg(O,H)$ (*top*), the Fe_3O_4 in MgO sample (*middle*), and α-Fe in MgO (*bottom*).

Figure 4 Hysteresis loops of the Fe_3O_4 in MgO sample (*dashed lines*), and α-Fe in MgO (*solid lines*).

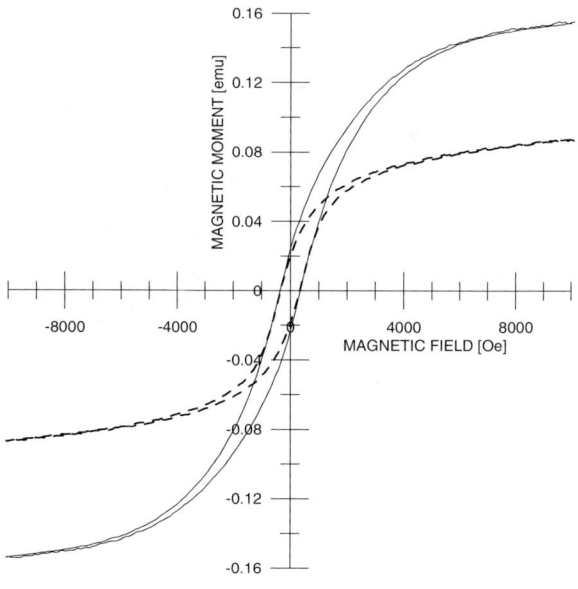

derived from this hysteresis loop is $H_C = 30 \pm 1$ mT. The phase composition of the nanocomposite remained unchanged during its heating in vacuum up to 820 K and magnetite was the only magnetic phase in the system.

The beginning of the reduction of magnetite allowing the preparation of the α-Fe in MgO nanocomposite was observed in the vicinity of the Curie temperature of magnetite. In this case annealing at 970 K was chosen. The mean size of iron particles synthesized by this way remained slightly above ($\sim$30 nm) the size of the original magnetite particles. The Mössbauer phase analysis (Figure 4) proved that the α-Fe dominates in the material but traces of other phases corresponding to Fe^{2+} can be analysed in the spectrum as well. Their total amount is smaller than 2% of iron atoms. They can be ascribed to the solid solution of iron atoms in MgO. The increase (2.3 times) in saturation magnetization in comparison with that containing the magnetite nanoparticles can be observed in the hysteresis loop of the samples of the nanocomposite. The increase corresponds well to the ratio of saturation magnetization of pure α-Fe to magnetite (values of coarse grains) which is approximately 2.1. The coercive force of iron nanoparticles $H_C = 30 \pm 1$ mT is only slightly lower than that for the nanocomposite with magnetite nanoparticles.

4. Conclusions

It is shown that Fe_3O_4 and α-Fe nanoparticles embedded in MgO matrix can be prepared by heat treatment of mixture of nanocrystalline Fe_2O_3 with Mg(H,O) in reduction atmosphere. Spherical Fe_3O_4 nanoparticles (mean size $\sim$20 nm) are ferromagnetic with $H_C = 25 \pm 1$ mT. The α-Fe particles ($d \sim 30$ nm) showed coercivity $H_C = 30 \pm 1$ mT.

Acknowledgments   The author (O.S.) thanks to the Organizing Committee for the financial support for participation in the 8th Solid State Physics Conference. This work was supported by the Grant Agency of the Czech Republic (Contract No. 202/04/0221).

References

1. Schmid, G.: In: Schmid, G. (ed.) Nanoparticles from theory to application, p. 1. Wiley, Berlin (2003)
2. Hui, S., Zhang, Y.D., Xiao, T.D., Wu, M., Ge, S., Hines, W.A., Budnick, J.I., Yacaman, M.J., Troiani, H.E.: In: Komarneni, S., Parker, J.C., Vaia, R.A., Lu, G.Q., Matsushita, J.-I. (eds.) Nanophase and nanocomposite materials IV, MRS Mat. Res. Soc. Symp. Proc. vol. 703, p. V3. Warrendale (2001)
3. Herzer, G.: IEEE Trans. Magn. **26**, 1397–1402 (1990)
4. Herzer, G.: J. Magn. Magn. Mater. **112**, 258–262 (1992)
5. Chin, G.Y., Wernick, J.H.: In: Wohlfarth, E.P. (ed.) Ferromagnetic materials, p. 55. North Holland, Amsterdam (1980)
6. Brabers, V.A.M.: In: Buschow, K.H.J. (ed.) Handbook of magnetic materials, vol. 8, p. 189. Elsevier, Amsterdam (1995)
7. Zbořil, R., Machala, L., Mašláň, M., Tuček, J., Müller, R., Schneeweiss, O.: Phys. Status Solidi (c) **1**, 3710–3716 (2004)
8. Schneeweiss, O., Jirásková, Y., Šebek, J.: Phys. Status Solidi A **189**, 725–729 (2002)

Hyperfine Interact (2005) 164: 41–49
DOI 10.1007/s10751-006-9231-7

How can Mössbauer Spectrometry Contribute to the Characterization of Nanocrystalline Alloys?

M. Miglierini

Abstract A controlled annealing of amorphous precursors represents a simple way of obtaining nanocrystalline alloys featuring interesting magnetic properties suitable for technical applications. They stem from the presence of crystalline nanograins embedded in the amorphous residual matrix which determine the resulting macroscopic parameters. In order to understand correlation between the microstructure and the resulting magnetic behaviour, Mössbauer spectrometry is used as a method of local probe analysis. Possibilities of this technique are discussed and representative examples of investigation of NANOPERM-type nanocrystalline alloys are provided.

Key words Mössbauer spectroscopy · hyperfine interactions · nanocrystalline alloys

1. Introduction

Nanocrystalline alloys exhibit interesting combination of high magnetic saturation and high magnetic permeability. Because of this and because of their promising practical applications, these alloys are attractive as objects of further examination. Nanocrystalline alloys are readily attained from amorphous precursors by a controlled annealing. Partial crystallization gives rise to creation of nanograins of the order of 5 to 30 nm. The suppression of the effective magnetic anisotropy and the decrease of magnetostriction are sources of their soft magnetic properties. Apart from composition, the amount and type of nanograins created during thermal treatment of the master alloy determine the resulting macroscopic parameters which are superior to conventional and/or amorphous alloys. Thus, the resulting magnetic properties of the alloy can be tailored to meet the particular needs suitable for technical applications.

M. Miglierini (✉)
Department of Nuclear Physics and Technology, Slovak University of Technology,
Ilkovicova 3, 812 19 Bratislava, Slovakia
e-mail: marcel.miglierini@stuba.sk

Figure 1 Transitions between
nuclear levels.

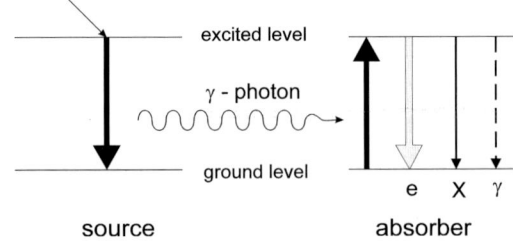

There are three main families of nanocrystalline alloys prepared from rapidly quenched ribbons: FINEMET [1], NANOPERM [2], and HITPERM [3] with the generalised composition of Fe–Nb–Cu–Si–B, Fe–M–(Cu)–B, and Fe,Co–M–(Cu)–B, correspondingly. All are characterised by an appearance of Fe-based crystalline phases during the first crystallization step. Whereas NANOPERM exhibits nearly pure bcc-Fe phase, presence of Si and Co in the other two alloys modifies the structural arrangement of Fe atoms by inclusions of the respective elements into the iron lattice positions.

In order to understand macroscopic magnetic properties of nanocrystalline alloys, microstructural arrangement comprising mutual interactions between the amorphous residual matrix and the crystalline phase should be known. Usually, X-ray diffraction, transmission electron microscopy, and differential scanning calorimetry are used to characterise the structural behaviour [4] and magnetic measurements help with the elucidation of magnetic parameters [5]. To describe the correlation between the microstructure and the resulting macroscopic magnetic properties the usage of techniques sensitive to local atomic arrangement is inevitable. Among them, Mössbauer spectrometry acquires a special importance due to the possibility to probe simultaneously structural characteristics and magnetic interactions [4, 5].

In the following, we demonstrate the diagnostic potential of Mössbauer effect techniques in the investigation of nanocrystalline alloys. Prior of doing that basic principles and techniques of the method are described.

2. Basic features of Mössbauer spectroscopy

Mössbauer spectroscopy [6] is one of the most sensitive tools for structural and magnetic investigations. Recoilless nuclear gamma resonance – the Mössbauer effect [7] – takes place between the same type of nuclei located in a source and an absorber (sample) as schematically drawn in Figure 1. Even though about 47 elements with almost 110 transitions allow the observation of the Mössbauer effect more than 65% of all scientific works are done using the isotope of iron-57. Because of this, the term *Mössbauer spectroscopy* often means ^{57}Fe *Mössbauer spectroscopy*.

Typical Mössbauer effect experiments are performed in transmission geometry, the so-called Transmission Mössbauer spectrometry (TMS). The whole sample is exposed to gamma radiation which is subsequently detected behind it. Hence, the selective resonance absorption probes the entire bulk of the specimen. During de-excitation of the sample, several types of radiation including (Figure 2): (1) Gamma photons, (2) X-rays, and (3) conversion electrons are emitted. Since the escape depth of conversion electrons is about 150 nm we can scan the immediate sample

Figure 2 Scattering techniques.

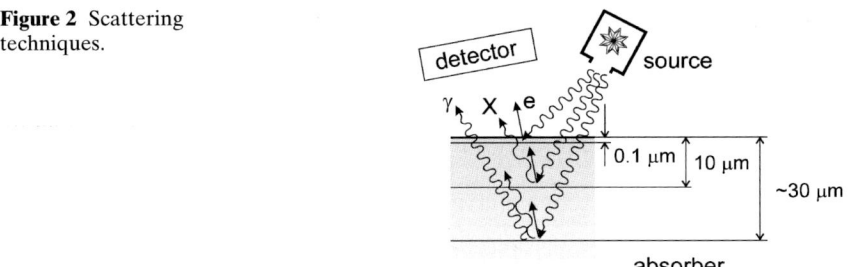

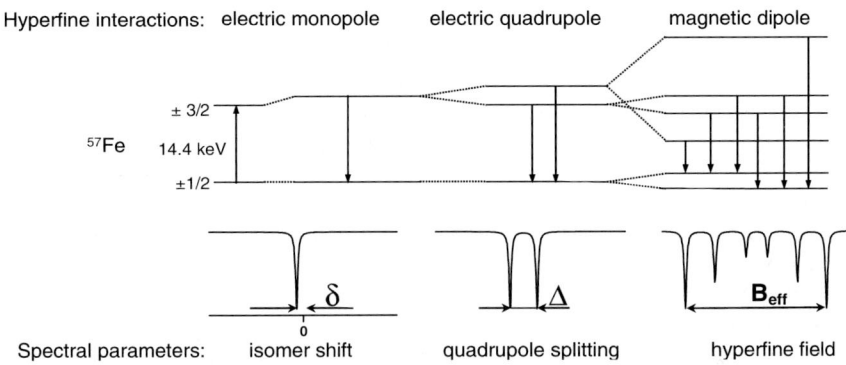

Figure 3 Hyperfine interactions in ^{57}Fe nuclei and their reflection in Mössbauer spectra.

surface. The associated Mössbauer effect technique is known as Conversion Electron Mössbauer Spectrometry (CEMS).

Atomic nuclei generally exhibit electric and/or magnetic moments, which causes the energetic levels to split as demonstrated in Figure 3. Respective spectral parameters provide information about intensities of hyperfine magnetic fields (B_{eff}) or electric field tensors (1), and/or electron densities at the nuclei (δ). Line widths give information on the structure (defects, order), and line intensities (line areas) correspond to the orientation of the sample's magnetization and/or determine the texture. Area under all lines of an individual subspectrum is directly proportional to the relative amount of resonant atoms located in that particular structural site.

Figure 4 illustrates how the structural distinctions between ordered (crystalline) and disordered (amorphous) arrangements of resonant atoms are reflected in Mössbauer spectra. Crystalline materials feature discrete values of spectral parameters derived from narrow and well-separated spectral lines. They are reflecting precise positions of resonant atoms in a crystalline lattice well-defined over several lattice constants. Distinct crystalline phases can be unambiguously identified by the help of spectral parameters which are like fingerprints for the crystalline arrangement. At the same time, doublet-like spectral pattern indicates that the investigated sample is non-magnetic whereas sextuplet of spectral lines characterises magnetic samples.

On the other hand, amorphous material exhibits broad and overlapped lines. This is caused by complete randomness of structural arrangement of the constituent atoms. As a result, each resonant atom experiences non-uniform chemical and

 Springer

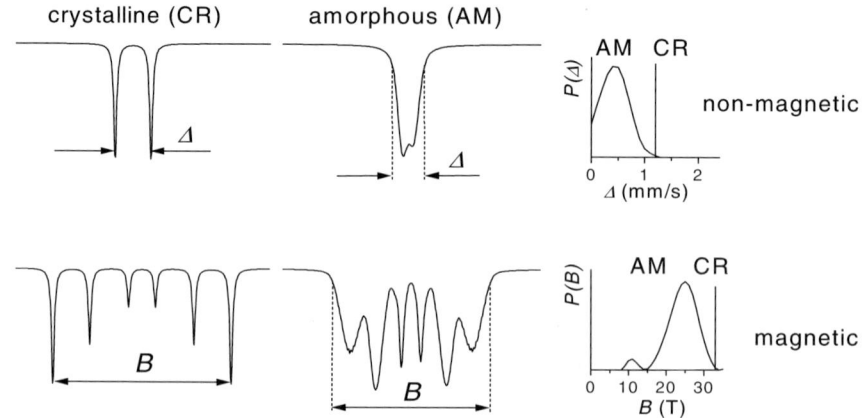

Figure 4 Mössbauer spectra of crystalline and amorphous materials in non-magnetic and magnetic states. *Right-hand panels* show corresponding spectral parameters.

topological sites occupied by different nearest neighbours. Consequently, instead of discrete values distributions of individual spectral parameters are observed. Nevertheless, distinction between magnetic and non-magnetic systems is also possible as in the case of crystalline samples.

Mössbauer spectrometry, that is complementary to diffraction techniques, is often proving decisive in the area of material science [8, 9]. Indeed, its local behaviour enables to be an atomic scale sensitive probe while its time of measurement allows investigation of relaxation phenomena and dynamic effects.

3. Mössbauer spectroscopy in NANOPERM-type nanocrystalline alloys

Due to their composition, NANOPERM alloys demonstrate improved macroscopic magnetic properties [2]. They are more interesting also for Mössbauer effect studies because they do not contain any Si. Consequently, nearly pure bcc-Fe nanograins emerge during the first crystallization. Since this is actually a calibration material for Mössbauer spectrometry and taking also more simple shape of Mössbauer spectra into account, NANOPERM-type alloys represent very suitable material for model case studies.

As mentioned above, nanocrystalline alloys are prepared from amorphous precursors. Lack of any periodicity in such highly disordered systems affects the appearance of the Mössbauer spectra in such a way that the corresponding lines are broad with a considerable overlap. They are analysed by the help of distributions of hyperfine parameters (hyperfine magnetic fields, B, for magnetically active samples, and quadrupole splitting, Δ, for the paramagnetic ones). The latter provide information on the short-range order, *i.e.* structural arrangement over few nearest neighbouring shells.

Presence of minute amounts (already about 1%–2%) of resonant atoms positioned in a crystalline lattice is indicated by well-distinguished narrow spectral lines. The derived parameters are single values that unambiguously identify the kind of the crystalline phase observed. Thus, the striking difference in line width as well as

 Springer

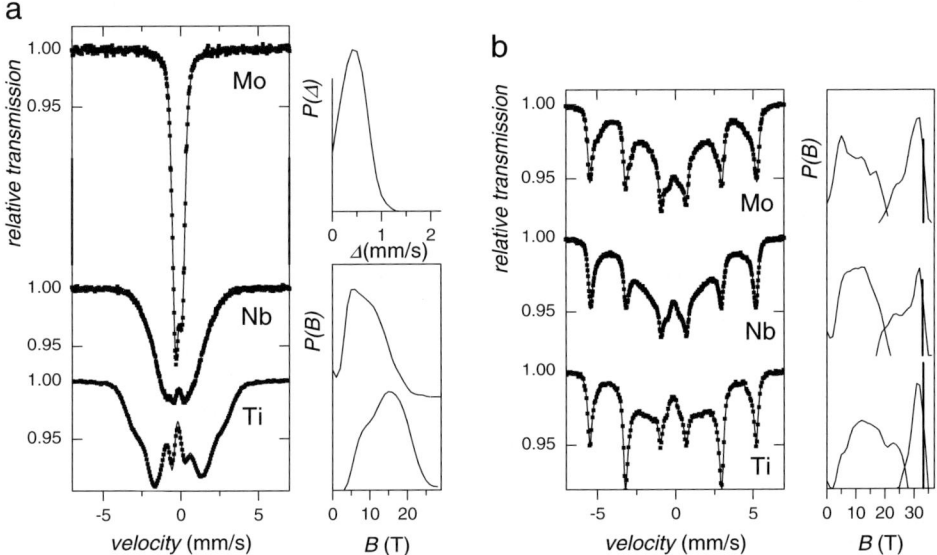

Figure 5 TMS spectra (*left-hand panels*) and corresponding distributions of hyperfine interactions (*right-hand panels*) of the $Fe_{80}M_7B_{12}Cu_1$ alloys in as-quenched (**a**) and nanocrystalline (**b**) states for $M = $ Mo, Nb, and Ti.

usually in line positions between amorphous and crystalline phases enable us to recognize the onset of formation and further progress of the crystallization process at the very first sight. Moreover, line areas are proportional to the relative ratio of the resonant atoms situated in disordered (amorphous) and ordered (crystalline) structural positions.

3.1. Composition

Room temperature Mössbauer spectra of amorphous $Fe_{80}M_7B_{12}Cu_1$ alloys are compared in Figure 5a with respect to the metal contents $M = $ Mo, Nb, and Ti. An influence of the sample's composition on their magnetic state is clearly demonstrated by the apparent qualitative differences in the shapes of the absorption lines. The Mo-containing amorphous alloy is paramagnetic at room temperature and its Mössbauer spectrum shows an asymmetric doublet. The $M = $ Nb sample depicts broad and unresolved Mössbauer lines which indicate that resonant atoms are already in magnetically ordered state. Finally, the $M = $ Ti amorphous alloy exhibits even more strong magnetic interactions at room temperature as the $M = $ Nb one.

Mössbauer spectra of nanocrystallized samples prepared from the amorphous precursors in Figure 5b consist of sharp Lorentzian sextuplets which correspond to crystalline phases, and of strongly broadened sextets which can be attributed to the retained amorphous phase. The amorphous subspectra for the $M = $ Nb and Ti nanocrystalline alloys are broader than those for $M = $ Mo suggesting that contributions of higher hyperfine magnetic fields appear. It should be noted that though the $M = $ Mo as-quenched alloy is paramagnetic at room temperature its

 Springer

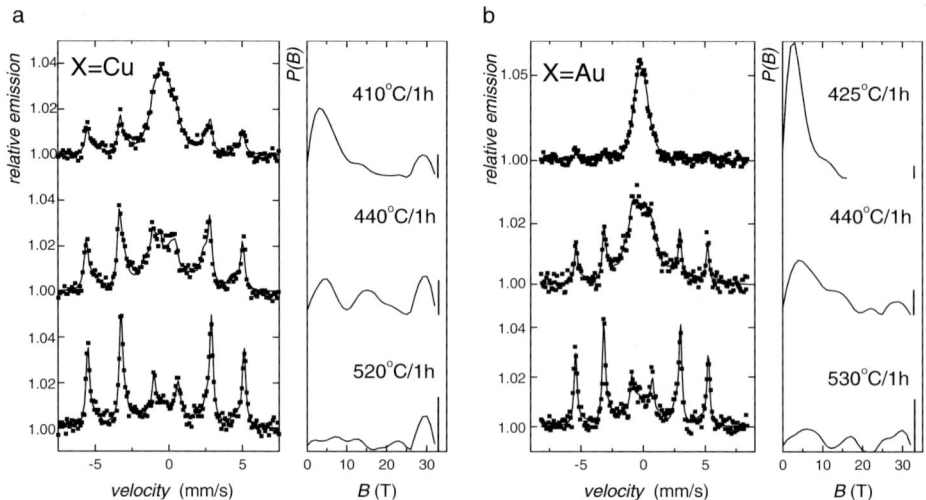

Figure 6 CEMS spectra (*left-hand panels*) and corresponding distributions of hyperfine magnetic fields, *P(B)* (*right-hand panels*) of the $Fe_{80}Mo_7X_1B_{12}$ alloy for **a** $X = $ Cu and **b** $X = $ Au annealed at the indicated temperatures for one hour.

nanocrystalline counterpart depicts rather strong magnetic interactions inside the retained amorphous phase.

The effect of varying the composition of the nucleation metal is shown in Figure 6a and b where CEMS spectra of $Fe_{80}Mo_7X_1B_{12}$, $X = $ Cu and Au, respectively, are given together with the corresponding hyperfine magnetic field distributions, $P(B)$. The thick vertical lines in the $P(B)$ panels represent hyperfine field values of the crystalline bcc phase. Copper forms crystallization centres and its replacement by Au, even though minute in amount, has a significant impact on the crystallization behaviour.

This is obvious at low annealing temperature where the surface crystallization in the $X = $ Cu is already well developed whereas for $X = $ Au it is just starting. The corresponding $P(B)$ exhibit higher probabilities of low hyperfine magnetic fields in the whole temperature range. The latter indicated presence of atomic regions inside the amorphous residual phase which are paramagnetic.

3.2. Amount of nanocrystallites

When the originally amorphous $Fe_{76}Mo_8Cu_1B_{15}$ metallic alloy was subjected to heat treatment at elevated temperatures a progress of crystallization was observed. After annealing at $t_a = 450°C$ first traces of bcc Fe were found. With rising t_a their contribution became more pronounced. Continuous transformation from purely amorphous into partially crystallised structure and the formation of crystalline grains in the sample can be followed by progressive appearance of narrow spectral lines in Mössbauer spectra as demonstrated in Figure 7. It should be noted that TMS spectra are displayed up-side-down for the sake of clarity of the 3D illustration.

Magnetic states of iron atoms are significantly affected by already small crystalline fraction. At room temperature (RT), a remarkable broadening of the central part of

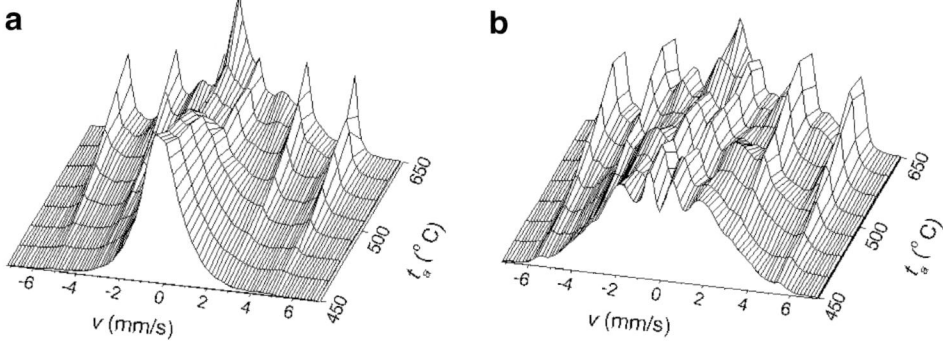

Figure 7 Mössbauer spectra of $Fe_{76}Mo_8Cu_1B_{15}$ metallic alloy taken at **a** RT and at **b** LNT from samples annealed at the indicated temperatures t_a for 1 h.

Mössbauer spectra in Figure 7a is observed around the velocity range of ± 2 mm/s towards higher crystalline contents, *i.e.* higher t_a. It originates from ferromagnetic interactions among the grains which are penetrating into the amorphous residue [10] thus strengthening the magnetic hyperfine fields within the sample.

Higher contribution of magnetic regions depends also on temperature of measurement. Mössbauer spectra in Figure 7b which have been taken at 77 K (LNT) demonstrate well-resolved six-line broad patterns in the central velocity region. At this temperature, the investigated alloy is far below its Curie temperature thus exhibiting more evident magnetic interactions. Narrow Mössbauer lines which represent newly formed bcc crystalline phase are also clearly visible.

3.3. Effect of temperature

Figure 8 shows Mössbauer spectra of nanocrystalline $Fe_{80}Mo_7Cu_1B_{12}$ alloy recorded at the indicated temperatures and their distributions of hyperfine fields $P(B)$.

By varying temperature of measurement, it is possible to separate contributions from different spectral components due to diversity in their magnetic states. This (1) simplifies the fitting procedure, and (2) opens a new approach in the investigation of hyperfine fields because different structural and/or magnetic components are better resolved. When the measuring temperature increases and becomes higher than the expected Curie temperature of the residual amorphous phase, the hyperfine structure changes drastically: The broad line magnetic sextet collapses to a paramagnetic doublet as demonstrated in Figure 8. It is important to emphasise that the evolution of Mössbauer spectra depends strongly on the volumetric fraction of the crystalline phase, which governs the nature and the strength of intergrain magnetic interactions and on exchange magnetic interactions within the crystalline grains and the residual amorphous matrix.

3.4. Surface modification

CEMS Mössbauer spectra of nanocrystalline $Fe_{76}Mo_8Cu_1B_{15}$ alloy after laser treatment with one pulse with the indicated fluences are shown in Figure 9. Laser treatment was performed in the atmosphere of nitrogen. Surface melting caused by the

Figure 8 (*Left*) TMS spectra
Mössbauer spectra of
$Fe_{80}Mo_7Cu_1B_{12}$ nanocrystalline alloy taken at the
indicated temperatures.

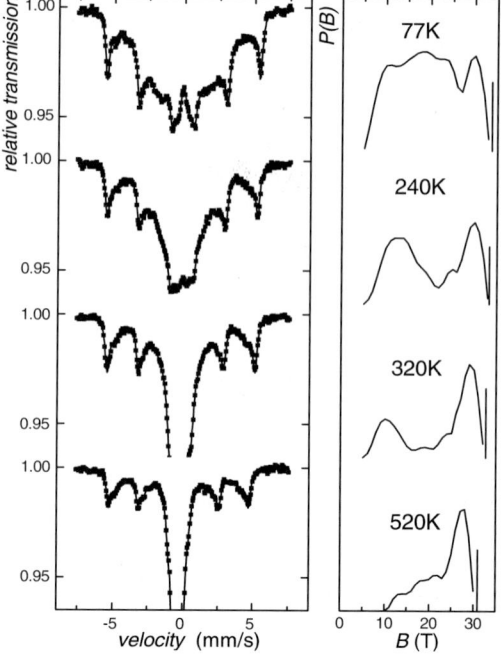

Figure 9 (*Right*) CEMS
spectra of the nanocrystalline
$Fe_{76}Mo_8Cu_1B_{15}$ alloy after
laser treatments with one pulse
with the indicated fluences.

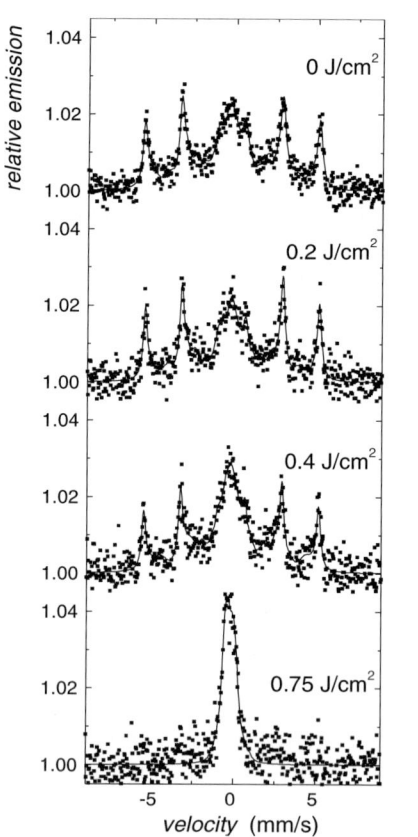

laser beam and subsequent rapid quenching due to the thermal contact with neighbouring regions of the sample's material have lead to complete re-amorphization after the irradiation with the fluence of 0.75 J cm^{-2}. As a result, the corresponding Mössbauer spectrum shows only broadened central doublet. The magnetic regions of the retained amorphous phase (comprising in this case 100% of the surface) have vanished entirely, too.

4. Conclusions

Mössbauer spectroscopy proves to be a valuable tool for the study of magnetic property-to-microstructure relationship for nanocrystalline alloys. In particular, the two techniques – transmission and detection of conversion electrons – allow differentiating between contributions from surface and bulk atoms to the overall magnetic behaviour of the samples studied. Using the results from Mössbauer spectroscopy, the magnetic structure and/or microstructure of nanocrystalline alloys can be modelled.

Acknowledgments The author is indebted to B. Idzikowski (Poznan), D. Janickovic (Bratislava), P. Schaaf (Göttingen), and J. M. Grenèche (Le Mans) for their involvement. This work was supported by the grants SGA 1/1014/04 and FR/SL/FEISTU/04.

References

1. Yoshizawa, Y., Oguma, S., Yamauchi, K.: J. Appl. Phys. **64**, 6044 (1988)
2. Suzuki, K., Kataoka, N., Inoue, A., et al.: Mater. Trans., JIM **31**, 743 (1990)
3. Willard, M.A., Laughlin, D.E., McHenry, M.E., et al.: J. Appl. Phys. **84**, 6773 (1988)
4. Miglierini, M., Kopcewicz, M., Idzikowski, B., et al.: J. Appl. Phys. **85**, 1014 (1999)
5. Hasiak, M., Miglierini, M., Ciurzyñska, W.H., et al.: Mater. Sci. Eng., A **375–377**, 1053 (2004)
6. Mössbauer, R.L.: Naturwissenschaften **45**, 538 (1958)
7. Mössbauer, R.L.: Z. Phys. **151**, 124 (1958)
8. Miglierini, M., Petridis, D. (eds.): Mössbauer Spectroscopy in Materials Science. Kluwer, Dordrecht, The Netherlands (1999)
9. Mashlan, M., Miglierini, M., Schaaf, P. (eds.): Materials Research in Atomic Scale by Mössbauer Spectroscopy. Kluwer, Dordrecht, The Netherlands (2003)
10. Navarro, I., Ortuño, M., Hernando, A.: Phys. Rev. **B53**, 11656 (1996)

Hyperfine Interact (2005) 164: 51–65
DOI 10.1007/s10751-006-9233-5

Nanostructure and Phases Formation under Mechanical Alloying of Bynary Powder Mixtures of Fe and sp-Element (M); M = C,B,Al,Si,Ge,Sn

E. P. Yelsukov · G. A. Dorofeev

Abstract The processes of mechanical alloying of iron and sp-elements (C, B, Al, Si, Ge, Sn) under identical conditions of mechanical treatment have been studied. General regularities and differences in the mechanisms and kinetics of solid state reactions have been ascertained. A microscopic model of mechanical alloying in these systems is suggested.

Key words mechanical alloying · iron and sp-elements · solid state reactions · mechanisms · kinetics.

1. Introduction

One of the main points in studying mechanical alloying (MA) is finding out microscopic mechanisms of solid state reactions (SSRs), in particular, those of forming supersaturated solid solutions. In other words, what do we imply by the term 'deformation atomic mixing'? It is also necessary to find out the major factors determining the SSRs kinetics. Appropriate model systems for such an investigation are binary powder mixtures of Fe with sp-elements (M) such as C, B, Al, Si, Ge, Sn. The ratio of the covalent radii R_M/R_{Fe} changes in a wide range from 0.66 (C) to 1.21 (Sn). The equilibrium phase diagrams of the Fe–M alloys are characterized by different types: With actual absence of solubility of the M atoms in α-Fe (Fe–C and Fe–B), with low solubility (Fe–Sn) and broad concentration range of solid solutions (Fe–Al, Fe–Si, Fe–Ge). MA in these Fe–M systems has been attracting much attention of many research teams for the last 15 years. A detailed analysis of the data published is presented in our papers [1–11]. In general, one can ascertain that MA is proceeded by the formation of laminar structure with characteristic sizes of 1–10 μm (see, e.g. [12, 13]). However, detailed comparison of the mechanisms and kinetics of MA in

E. P. Yelsukov (✉) · G. A. Dorofeev
Physical-Technical Institute UrB RAS, 426001 Izhevsk, Russia
e-mail: yelsukov@fnms.fti.udm.ru

the Fe–M systems on the basis of the earlier published data is not possible for the following reasons:

a.   MA was carried out under different conditions: The material of grinding tools, power intensity of milling devices. The latter characteristic has not been presented as a rule;
b.   There are considerable differences in the results published.

In the present work we have classified the results of our investigation of the mechanisms and kinetics of MA in the Fe–M systems under equal conditions of treatment in a mill with the known power intensity and controlled levels of contamination and heating of the samples studied. The results of other authors have been taken into account as well.

2. Experimental

For MA the mixtures of pure Fe (99.99) and sp-elements (99.99) powders with the particle size less than 300 μm were taken. MA was carried out in an inert atmosphere (Ar) in a planetary ball mill Fritsch P-7 with the power intensity 2.0 W/g. For each given mechanical treatment time the mass of the loaded sample was 10 g. With the given power intensity the time of mechanical treatment of 1 h corresponds to the dose of 7 kJ/g. Using air force-cooling, the heating of the vials, balls and sample did not exceed 60°C. The milling tools – vials (volume 45 cm^3) and balls (20, diameter 10 mm) were made of hardened steel containing 1 wt.% C and 1.5 wt.% Cr. Possible getting of the milling tools material to the sample was monitored by the measurement of the powder, vial and ball mass before and after treatment. X-ray diffractometer with Cu K$_\alpha$ monochromatized radiation was used for X-ray examinations of samples. Phase composition was determined from X-ray diffraction (XRD) patterns using Rietveld method. The crystallite size ($<L>$) and microstrains ($<\varepsilon^2>^{1/2}$) were calculated from peak profile analysis using the Voigt function. Room-temperature Mössbauer investigations were carried out with a conventional constant acceleration spectrometer and ^{57}Co(Cr) source. To estimate the size of the powder particles after MA an Auger spectrometer in secondary-electron microscopy mode image was used. For all the systems studied the particles size after MA was in the range of 5–20 μm.

3. Results and discussion

3.1. Milling of pure Fe

Under the given conditions we showed in [14] that with the milling time $t_{mil} = 1$ h the grain size in α-Fe particles $<L> = 13$ nm. On increasing the t_{mil} up to 16 h $<L>$ goes down to 9 nm, the bcc lattice parameter of α-Fe increases up to 0.2869 nm. In [14] the concept of interface regions is introduced, which include the boundary and close-to-boundary distorted zones. The width of interfaces (d) can be estimated, according to [15, 16], in 1 nm. With $<L> = 9$ nm the volume fraction of interfaces (f_{if}) is ~15%. Supposing that in the body of the grain the bcc lattice parameter is equal to 0.2866 nm, we obtain the average lattice parameter in distorted zone of interfaces

 Springer

0.2887 nm, i.e. almost by 1% more than in typical α-Fe. The existence of the distorted zones is confirmed by symmetrical broadening by 20% of the lines of the Mössbauer spectrum (MS) of the milled α-Fe without emergence of any new components in the spectrum.

3.2. Mechanical alloying of Fe-C and Fe-B

Carbon and boron practically do not dissolve in α-Fe, their covalent radii are much less than that of Fe: $R_C/R_{Fe} = 0.66$ and $R_B/R_{Fe} = 0.75$. It was established [7, 8, 12, 17] that in the C concentration range $0 < x \leq 25$ at.% in the initial mixture there was a critical concentration $x = 17$ at.% at which the change of the type of SSRs in MA took place. In [5, 8] we found the following types of SSRs with $x \geq 17$ at.% C.

$$17 \leq x < 25, \; Fe + C \rightarrow Fe + Am(Fe - C) \rightarrow Fe + Am(Fe - C) + (Fe_3C)_D;$$
$$x = 25, \; Fe + C \rightarrow Fe + Am(Fe - C) \rightarrow (Fe_3C)_D,$$

where Am(Fe–C) designates the amorphous Fe–C phase the concentration of C in which is close to 25 at.%; $(Fe_3C)_D$ designates distorted cementite. All of these SSRs take place after reaching the nanocrystalline state in the α−Fe particles ($<L><$ 10 nm). The estimates of the Am(Fe–C) phase amount showed [5–8] that it formed in the interfaces of the α-Fe nanostructure. The given SSRs types with $x \geq 17$ agree with the data published earlier in [12, 18, 19] but they differ from the results of [17, 20] in which alongside with cementite hexagonal carbides were found and from [21, 22] in which only cementite as the first MA stage was found to be formed. Controversial results were obtained for a single-stage process of MA as well with $x < 17$ at.% C: Hexagonal carbides [17], interstitial solid solution of [23] and amorphous Fe–C phase [7, 8, 11]. Consider in detail the process of MA with the C concentration in the initial mixture $x = 15$ at.% [11]. In the XRD patterns, broadened bcc reflections are revealed, which positions do not change as t_{mil} increases and correspond to those of the α-Fe lines. The grain size decreases from 100 nm for the initial powder to 6 nm at $t_{mil} = 1$ h and 3 nm at $t_{mil} = 16$ h. Besides at the base of (110) peak there is a considerable increase of intensity, which is similar to the contribution from the first peak of the amorphous phase. In the MS and hyperfine magnetic field (HFMF) distribution functions $P(H)$ besides the component attributed to pure α-Fe ($H = 330$ kOe) there is a component with broad distribution of HFMF from 50 to 330 kOe. It is obvious that the given component corresponds to the Am(Fe–C) phase which produced 'halo' in XRD patterns. The known concentration dependence of the average value of HFMF for amorphous Fe–C films [24] allowed us to estimate the C concentration in the Am(Fe–C) phase as equal to ~25 at.% C. The average grain size calculated from XRD patterns and the amount of the amorphous Am(Fe–C) phase (f_{Am}) in the process of MA calculated from the MS are given in Figure 1. Comparing the dependences $f_{Am}(t_{mil})$ and $<L>(t_{mil})$ one can draw the conclusion that SSR of the amorphous phase formation takes place on condition of nanocrystalline state realization in the α-Fe particles. The atomic fraction of the Am(Fe–C) phase increases from 0.15 ($t_{mil} = 1$ h) to 0.67 ($t_{mil} = 16$ h). To account for the amorphous phase amount, the conception on the existence of interfaces is applied. Figure 1(b) gives the calculated dependence of the volume fraction of the interfaces $f_{if}(t_{mil})$ according to the obtained grain sizes (Figure 1(a)), the given interface width of 1 nm and on the assumption of a cubic grain shape. The comparison of the dependences $f_{Am}(t_{mil})$ and $f_{if}(t_{mil})$ illustrates not only qualitative but also quantitative agreement. It

Figure 1 Time dependences of the α-Fe grain size – (a), atomic fraction of amorphous phase (f_{Am}) and volume fraction of interfaces (f_{if}) – (b) during MA in the Fe(85)C(15) system [11].

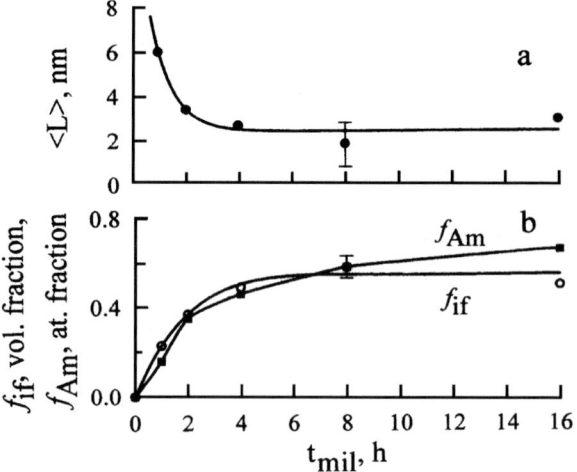

allows us to draw the conclusion on the amorphous phase formation in the interfaces of the α-Fe nanostructure. In Section 3.1 we estimated the bcc parameter of the distorted structure in the interfaces of 0.2887 nm. One can suppose that the increase of the size of the interstices in the distorted structure in comparison with that of the grain body of α-Fe ($a = 0.2866$ nm) makes the dissolution of C atoms in the interfaces easier under pulse mechanical treatment. Additional distortions during C dissolution lead to amorphization of the interfaces.

MA in the Fe–B system was studied in [25–30] with the B content in the initial mixture from 20 to 60 at.%. In [25] it was shown that with the content of B of 20 at.% MA was realized during one stage: Fe + B → Fe + Am(Fe–B). At the same time the stage of the amorphous phase formation precedes the formation of borides with a higher amount of B [26–30]. The given data show considerable similarity in the type of SSRs in the Fe–C and Fe–B systems. However, so far, detailed investigations of the kinetics of the initial SSRs stage in MA in the Fe–B system and its comparison with that of the Fe–C system have not been carried out. With this purpose in the present paper we have chosen the content of the initial mixture as Fe(85)B(15), mechanical treatment of which was carried out in the same conditions as for the Fe(85)C(15) mixture. All the changes observed in XRD patterns and MS of the mechanically alloyed samples Fe(85)B(15) are similar to those of the Fe(85)C(15) samples. The fractions of Fe atoms in the amorphous phase found from the MS depending on t_{mil} for Fe(85)B(15) (present work) and Fe(85)C(15) [11] are given in Figure 2(a). In both systems the condition of SSRs proceeding is reaching the nanostructure state in α-Fe (Figure 2(b)). However, in the Fe–C system the formation of the amorphous phase starts already at $t_{mil} = 1$ h. In the Fe–B system intensive growth of the amorphous phase amount is revealed at $t_{mil} > 4$ h and the maximum fraction of Fe atoms in it at $t_{mil} = 16$ h is 30%, i.e. almost twice as little as in the Fe–C system. The differences in the MA kinetics are revealed in the rate of the grain size decrease (Figure 2(b)) and in the level of microstrains (Figure 2(c)) as well. What is the reason of differences in kinetics of SSRs? Consider the kinetics of penetration and segregation of C and B atoms in the grain boundaries of α-Fe. For this purpose it is necessary to estimate what amount of B and C out of initial 15 at.% is chemically bound with the Fe atoms in the amorphous phase after MA ($x_{Am}{}^{B}$ and $x_{Am}{}^{C}$) and after the

 Springer

Figure 2 Comparative analysis of MA in Fe(85)C(15) [11] and Fe(85)B(15) (present work) mixtures: Fe atomic fraction in amorphous phase – (a), α-Fe grain size ($<L>$) – (b), α-Fe root mean-squared strain ($<\varepsilon^2>^{1/2}$) – (c), C and B amount in segregations on the α-Fe grain boundaries – (d).

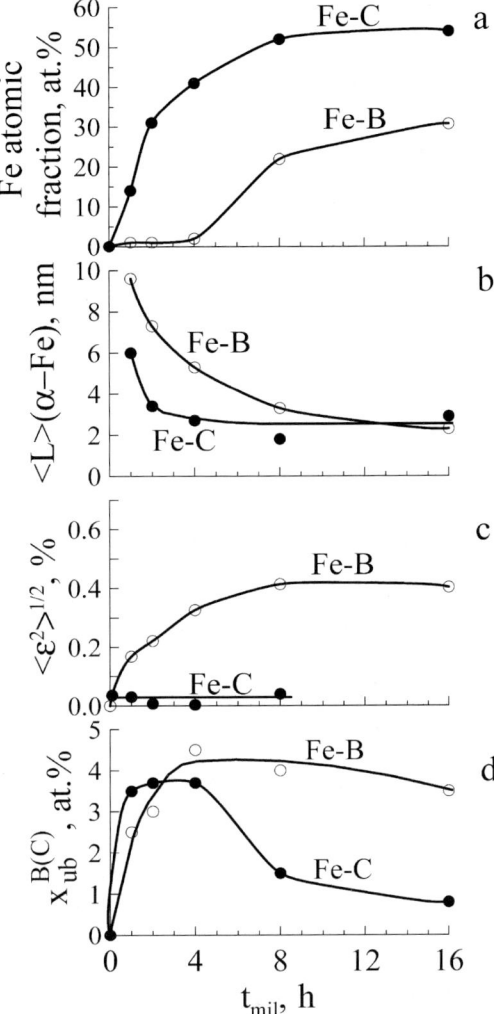

annealing following the MA ($x_{400}{}^B$ and $x_{500}{}^C$) in borides and carbides correspondingly (the subscripts indicate the annealing temperatures). The values of $x_{Am}{}^C$ and $x_{Am}{}^B$ can be found according to the known average values of HFMF $\overline{H}$ for the amorphous Fe–C and Fe–B alloys [24, 31] and from the known fraction of the Fe atoms in the amorphous phase (Figure 2,a). From the distribution functions of HFMF $P(H)$ for the mechanically alloyed samples Fe(85)C(15) and Fe(85)B(15) the values of $\overline{H}$ were calculated for the part of $P(H)$ function corresponding to the amorphous phase. From [24, 31] we found the maximum values of the C and B concentrations in the amorphous phases 25 and 20 at.%, respectively. Then, the maximum amount of C and B chemically bound with the Fe atoms $x_{Am}{}^C$ and $x_{Am}{}^B$ were calculated. Further procedure was to carry out a low-temperature annealing of the mechanically alloyed samples during 1 h at 500°C for the Fe(85)C(15) and at 400°C for the Fe(85)B(15). First, it was established that annealing at these temperatures of initial mixtures without mechanical treatment does not result in formation of any

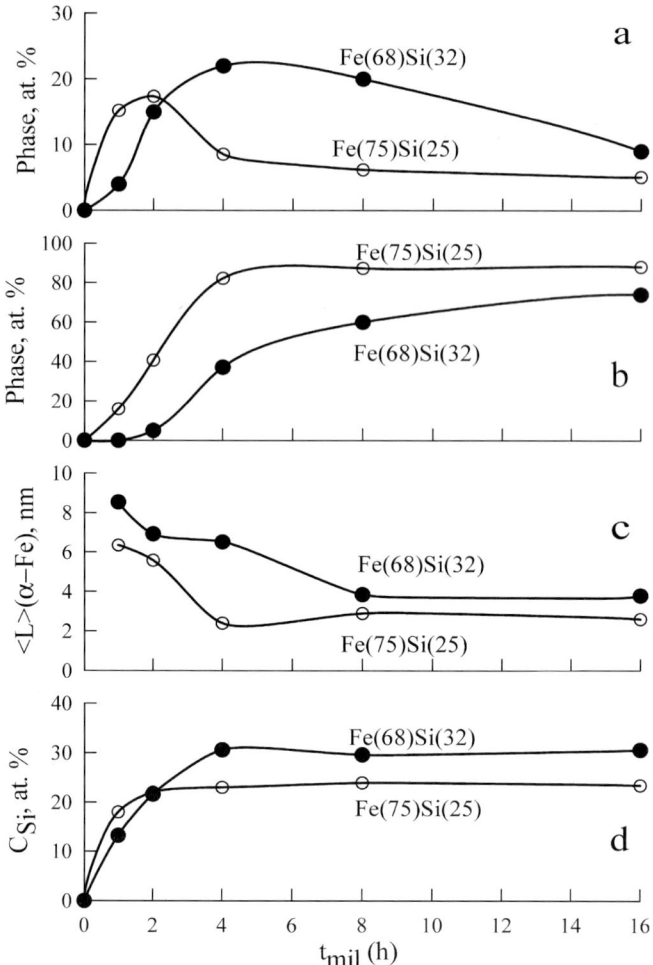

Figure 3 Comparative analysis of MA in Fe(68)Si(32) [4] and Fe(75)Si(25) [5] mixtures: ε-FeSi atomic fraction – (**a**), α-Fe(Si) atomic fraction – (**b**), α-Fe grain size – (**c**) and Si concentration in α-Fe(Si) solid solution – (**d**).

phases. Annealing of the samples after MA leads to the formation of the Fe_3C carbide in the Fe(85)C(15), $(Fe_2B)'$ and Fe_3B borides in the Fe(85)B(15). $(Fe_2B)'$ boride designated in [25] as x-Fe_2B is a metastable modification of the Fe_2B boride [28, 32]. From the MS of the annealed samples the fractions of the area referring to the carbide and borides were found. Then from the stoichiometric ratios in these phases the total amount of C ($x_{500}{}^C$) and B ($x_{400}{}^B$) chemically bound in the carbide and borides was calculated. The values $x_{ub}{}^C = x_{500}{}^C - x_{Am}{}^C$ and $x_{ub}{}^B = x_{400}{}^B - x_{Am}{}^B$ are the amount of the C and B atoms chemically unbounded with the Fe atoms in α-Fe particles after MA (Figure 2). The segregations of C and B atoms are at the α-Fe boundaries and they are the source for the amorphous phase formation in the interfaces of the α–Fe nanostructure. This is confirmed by the given above analysis of the Am(Fe–C) amount. From the presented in Figure 2(d) dependences $x_{ub}{}^C (t_{mil})$

 Springer

and $x_{ub}{}^B(t_{mil})$ it follows that the kinetics of the segregation formation at MA of the Fe(85)C(15) and Fe(85)B(15) systems coincides both in the time of mechanical treatment and by the amount of B and C in the segregations (~4 at.%). Under these conditions a considerable difference found in the kinetics of the amorphous phase formation (Figure 2(a)) can be accounted for by the difference in the covalent radii and, accordingly, by a different penetrating ability of B and C from the segregation into the close-to-boundary distorted zones.

3.3. Mechanical alloying of Fe–Si mixtures

The published data [4, 5, 33–38] carry inference that the most stable intermetallic compound (ε-FeSi) is formed at the first stage. Up to the concentration of 32 at.% Si [4, 5, 33], the supersaturated solid solution (SSS) is formed at the second stage. The process of MA in the Fe(68)Si(32) and Fe(75)Si(25) mixtures is given in detail in [4, 5]. The quantitative analysis of the MA process in these systems is illustrated in Figure 3, from which it is seen that:

a. All SSRs take place on α-Fe reaching the nanocrystalline state;
b. The kinetics of SSRs is conditioned by the concentration of Si in the initial mixture;
c. The maximum Si concentration in α-Fe(Si) SSS is realized practically simultaneously with its formation.

The amount of the ε-FeSi phase at the first stage of MA can be explained by its formation only in the interfaces of the nanostructure of the α-Fe particles. Using the values of the grain sizes in α-Fe particles $<L> = 5.5$ nm (Fe(75)Si(25), $t_{mil} = 2$ h), $<L> = 6.5$ nm (Fe(68)Si(32), $t_{mil} = 4$ h) and the width of the interfaces $d = 1$ nm, we obtain the volume fractions of the interfaces $f_{if} = 30$ and 20%, which agree with the maximum amounts of the ε-FeSi phase satisfactorily.

3.4. Comparative analysis of mechanical alloying in the Fe–M systems (M is Si, Ge and Sn isoelectron sp-elements)

The ratio of covalent radii R_M/R_{Fe} is 0.95, 1.04 and 1.21 for Si, Ge and Sn, respectively. The Fe–Si and Fe–Ge systems have a rather extended range of equilibrium solid solutions, meanwhile the solubility of Sn in α-Fe is low (3.2 at.% at 600°C). The sequence of SSRs in the Fe–Ge and Fe–Sn systems is similar to that in the Fe–Si system. At the first stage FeGe$_2$ intermetallic in the amorphous or nanocrystalline modifications is formed as well as FeSn$_2$, at the second stage – supersaturated solid solutions of Ge and Sn in α-Fe if the Ge and Sn concentration in the initial mixtures does not exceed 32 at.% [1–3, 9, 10, 33, 38–44]. It is of interest to compare the dependences of the phase amount, Si (Ge, Sn) concentrations in SSS and structural parameters for the given systems on the milling time in MA. Figure 4(a) and (b) collect the time dependences of the compounds and SSS amount; Figure 4(c) shows the α-Fe grain size dependences $<L>$ (t_{mil}). Their comparison shows that all SSRs take place on α-Fe reaching a nanocrystalline state, a less grain size being necessary to form SSS than to form an intermetallic compound. For all the systems an average grain size in SSS of α-Fe(M) is 2–4 nm. In [4, 5, 9] we showed that the amount of the intermetallic formed at the first stage of MA can be explained by its formation in the interface regions of the nanocrystalline α-Fe. Thus, the first stage of MA includes

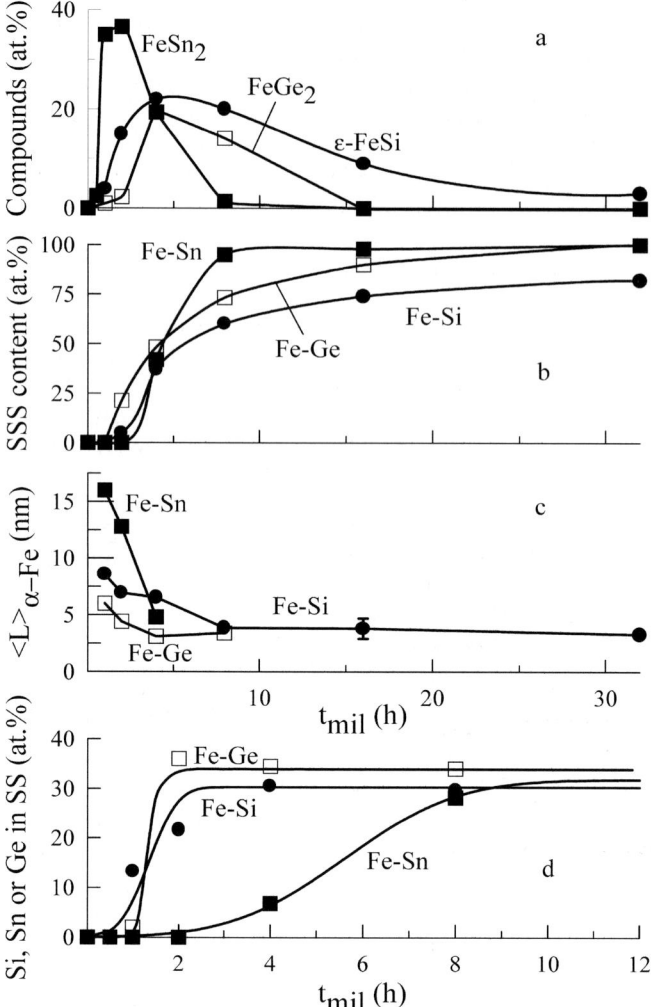

Figure 4 Comparative analysis of MA in Fe(68)M(32); M = Si, Ge and Sn systems [6]: intermetallic compound atomic fraction – (**a**), α–Fe(M) atomic fraction – (**b**), α-Fe grain size – (**c**) and M concentration in α-Fe(M) solid solution – (**d**).

penetration of sp-atoms along the grain boundaries of α-Fe, their segregation at the boundaries and formation of the first Fe–M phases in the interfaces. Formation of the Si, Ge, Sn and Al segregation is considered in Section 3.6 of this paper in detail. The sp-element type influences the kinetics. Consider two extreme cases – MA in the Fe–Si and Fe–Sn systems. The $FeSn_2$ is formed and disappears fast, meanwhile the rate of the ε-FeSi formation is slower, and it is present during the whole MA process (Figure 4(a)). As for SSS, it is also formed much faster in the Fe–Sn system than in the Fe–Si one (Figure 4(b)). However, a different situation takes place in the saturation of solid solutions (Figure 4(d)). In the Fe–Si and Fe–Ge systems the maximum Si(Ge) concentration reaches actually simultaneously with the formation of the solid solution, meanwhile in the Fe–Sn system the solid solution is saturated

 Springer

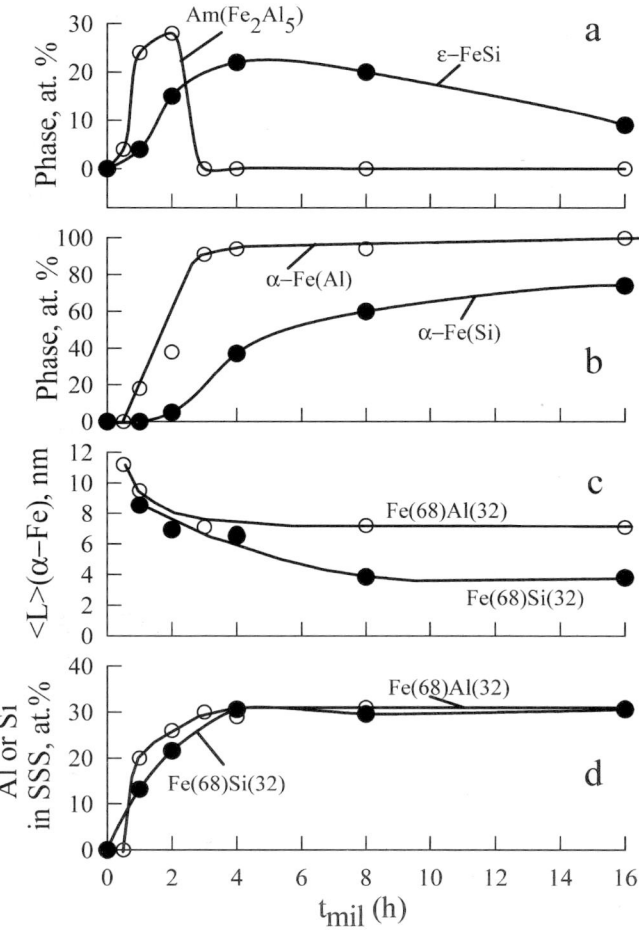

Figure 5 Comparative analysis of MA in Fe(68)Si(32) [6] and Fe(68)Al(32) (present work) mixtures: intermetallic compound atomic fraction – (a), α–Fe(Si) atomic fraction – (b), α-Fe grain size – (c) and Si(Al) concentration in α-Fe(Si, Al) solid solution – (d).

with Sn gradually. Concerning saturation of solid solution with Si, Ge and Sn one can make following supposition. At present, two possible mechanisms of accelerated diffusion in MA are discussed. There is interstitial diffusion at the collision moment [45] and diffusion along dislocations [46]. Taking into account the ratio of the atomic sizes in the Fe–Si, Fe–Ge and Fe–Sn systems, one can suppose that the accelerated diffusion in the Fe–Si and Fe–Ge systems proceeds mainly along interstices, but in the Fe–Sn system dislocation transfer takes place. Since the density of interstices is always higher than that of dislocations, the saturation rate of the solid solution in the Fe–Si system should be higher than in the Fe–Sn system. However, it is known [15] that with the grain size $<L> < 10$ nm there are no dislocations in the grain bulk. We consider the dislocation transfer as generation of dislocations at the moment of impact, their passing through the grain body and leaving for the boundary. Such a process can provide the penetration of the second component (Sn) into the grain bulk of α-Fe.

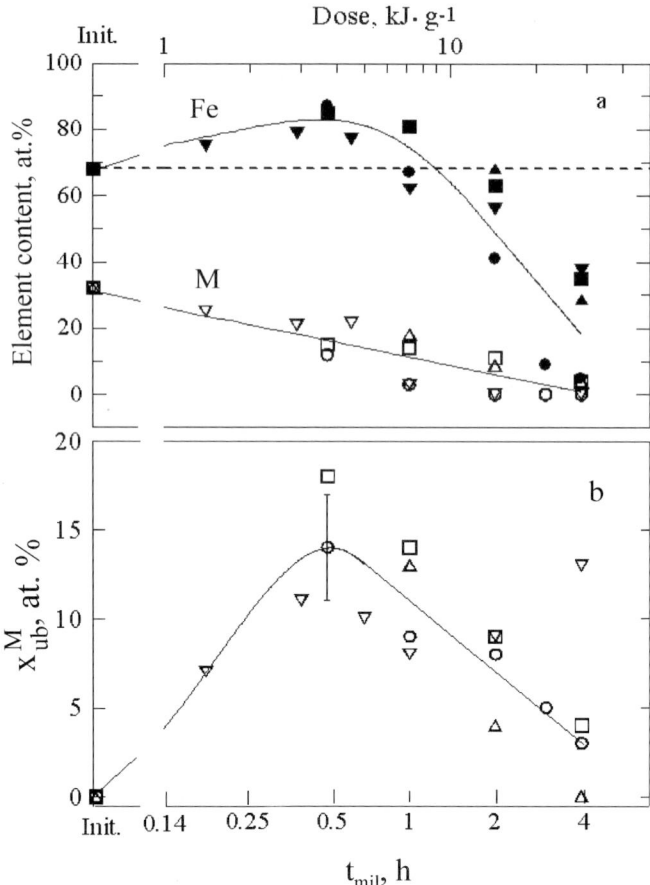

Figure 6 Contents of pure Fe (*closed symbols*) and M (*open symbols*) elements according to XRD data – (a) and calculated M-element amount (x_{ub}^M) in segregation against milling time (dose loaded mechanical energy) – (b)M = Al ($\bigcirc$), Si ($\square$), Ge ($\triangle$), Sn ($\triangledown$).

3.5. Comparative analysis of MA in the Fe–Al and Fe–Si systems

The Al and Si atoms have the sizes close to the Fe atom: $R_{Si}/R_{Fe} = 0.95$ and $R_{Al}/R_{Fe} = 1.01$. Equilibrium phase diagrams are characterized by the extended concentration range of solid solutions. The most important differences are the electron configuration of external shell of atoms, structure, melting temperature and mechanical properties of pure elements. MA in the Fe–Al system with the Al content in the initial mixtures x from 10 to 90 at.% has been intensively studied for the past 15 years [33, 47–67]. In most detail the type and mechanisms of SSRs were investigated for the mixtures with $x \geq 50$ at.% Al [47–53, 55 56] and close to the stoichiometric composition Fe(75)Al(25) [33, 52, 54, 56, 58, 64, 67]. The analysis of the published results leads to the following conclusions:

a.   MA takes place in the conditions on reaching the nanostructure state ($<L> \leq$ 10 nm);

 Springer

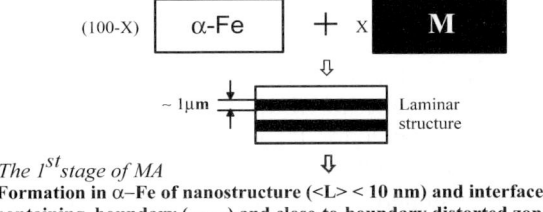

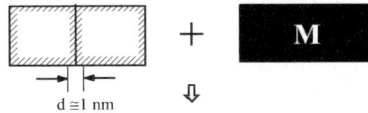

The 1ˢᵗ stage of MA
1. **Formation in α–Fe of nanostructure (<L> < 10 nm) and interfaces containing boundary (──) and close-to-boundary distorted zone (▨).**

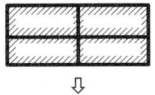

2. **Penetration of M atoms along grain boundaries, their segregation (━━) and decreasing <L>**

3. **Formation of amorphous Fe-C(B) phases with 20-25 at.% C(B) and Am(Fe₂Al₅), ε– FeSi, Am(FeGe₂), FeSn₂ intermetallic compounds with amorphous or nanocrystalline structures in interfaces (▨)**

The 2ⁿᵈ stage
4. **Formation of the Fe₃C and Fe₇C₃ carbides, Fe₂B boride if 15<x≤32 at.% C(B), α–Fe(M) SSS if x≤32 at.% Si(Ge, Sn) and x≤60 at.% Al.**

Figure 7 Microscopic model of mechanical alloying of Fe with sp-element (M); M = B, C, Al, Si, Ge and Sn.

b.   At the first stage of MA a phase is formed, which is revealed in the MS as a paramagnetic doublet at room temperature [48–50, 53, 55, 58, 64, 66]. With $x = 75$–80 at.% Al this phase is a final product of MA and is in the amorphous state [47–50, 53, 59, 60]. With $x \leq 50$ at.% Al it is found in the MS. In [53, 63, 64] it is supposed that this phase is due to the Fe diffusion into Al. The phase contains 70–80 at.% of Al, as in annealing it is transformed into intermetallic Fe_2Al_5 [51, 52, 56, 60, 61]. Further this phase will be referred to as $Am(Fe_2Al_5)$;

c.   It is established that the formation of the SSS α-Fe(Al) as a final product of MA takes place with $x \leq 60$ at.% Al though in a number of papers the formation of SSS was found with $x = 75$ at.% Al [55–57];

d.   It was shown [58, 64] that the composition of SSS from the very beginning of its formation is close to that of the initial mixture;

e.   Comparative study of the mechanisms and kinetics of SSRs in the Fe–Al and Fe–Si systems with equal compositions of initial mixtures and conditions of mechanical treatment has not been carried out yet.

   To compare the processes of MA in the Fe–Si and Fe–Al systems in the present work the composition Fe(68)Al(32) has been chosen. The XRD and MS data for mechanically alloyed samples Fe(68)Al(32) on the whole agree with the given above results of the published papers. However, in contrast to [53, 63, 64] we have not found any changes of the fcc lattice parameter of Al, that would point to the Fe solubility in Al. The amount of the phases formed, the α-Fe grain size and Al concentration in SSS α-Fe(Al) calculated from the XRD and Mössbauer data are presented in Figure 5. The latter was calculated according to [68, 69] from the values of the bcc lattice parameter and average values of HFMF for α-Fe(Al) SSS. The amount of the Am(Fe$_2$Al$_5$) phase can be accounted for its formation in the α-Fe particles interfaces. Under the given values of the undistorted part of the grain of α-Fe$<L>$ = 7 nm (t_{mil} = 2 h) and d = 1 nm we obtained the volume fraction of the interfaces f_{if} = 23%, correlating with the atomic fraction of this phase 29% (Figure 5 (a)). In Figure 5 the data for the Fe(68)Si(32) system are also presented. Comparing the results, we arrive at the conclusion that with a similar character of SSRs the kinetics of the phase formation in the Fe–Al and Fe–Si systems is substantially different. The amorphous phase Am(Fe$_2$Al$_5$) forms and disappears more quickly in comparison with the ε-FeSi phase (Figure 5(a)). The rate of SSS formation in the Fe–Al also exceeds that of the Fe–Si (Figure 5(b)), despite the fact that $<L>_{\alpha\text{-Fe(Al)}} > <L>_{\alpha\text{-Fe(Si)}}$ (Figure 5(c)). However, the kinetics of saturation of the SSS with Al and Si is practically similar (Figure 5(d)), which testifies to the interstitial character of the Al diffusion into the α-Fe grain under pulse mechanical treatment. The kinetics of the phase formation can be influenced by both individual properties of pure materials (mechanical properties, melting temperature, etc.) and atomic characteristics of sp-elements (atomic size and configuration of the external electron shell). To find out the main reason, consider the process of formation of Al, Si, Ge and Sn segregations at the initial stage of MA.

3.6. Al, Si, Ge and Sn segregation at the grain boundaries of the α-Fe nanostructure

The possibility of segregation of Al, Si, Ge and Sn atoms in the α-Fe particles in MA follows from the early disappearance of reflections from pure sp-elements in the XRD. Figure 6 (a) presents the dependences of the amounts of pure elements Fe, Al, Si, Ge and Sn on the milling time, obtained only from XRD. The increase of the α-Fe amount at low values of t_{mil} points either to sp-atoms leaving the coherent scattering regions, or to decreasing pure sp-element due to their interaction with the walls of the vial and balls, or due to the appearance of the vial and balls (hardened steel) material in the samples because of wear. In [10, 11] we showed for the Fe(50)Ge(50) mixture that the increase of the α-Fe amount according to the data of XRD can be explained only by the Ge atoms leaving the coherent scattering regions for the grain boundaries of α-Fe and their segregation at them as the chemical analysis showed the composition 50:50 in the mechanically alloyed samples, and no changes were found out in the masses of the sample, vial and balls before and after milling. The latter also holds for all the systems of Fe–M at $t_{mil} \leq 8$ h. Write the amount of atoms of sp-elements in the segregated state in the α-Fe particles, $x_{ub}(t_{mil})$, as: $x_{ub}(t_{mil}) = x(0) - [x(t_{mil}) + x_{IC}(t_{mil}) + x_{SSS}(t_{mil})]$, where $x(0)$ and $x(t_{mil})$ are the amount of sp-elements in the initial mixture and with the given value of t_{mil} according to the data of XRD, $x_{IC}(t_{mil})$ and $x_{SSS}(t_{mil})$ are the concentrations of sp-atoms, chemically bound with the Fe atoms in the intermetallic compound and supersaturated solid solution, determined by both the data of XRD and MS. The calculated values of x_{ub} for Al, Si, Ge and Sn are given in Figure 6(b), from which it is seen that independently of individual properties of pure materials the kinetics of segregations formation at $t_{mil} \leq 2$ h is similar. It means that individual characteristics of sp-atoms affect the kinetics of the phase formation. Retaining a large amount of segregated Sn atoms at t_{mil} = 4 h is a consequence of slow saturation of SSS with Sn, having a large covalent radius.

 Springer

Taking into account the segregation of sp-atoms at the grain boundaries of α-Fe, we have shown in the thermodynamic calculations the energy gain of SSS formation when the grain size becomes lower than some critical one [3]. Besides, the presence of segregations allows to understand the phase formation in the interfaces of the α-Fe particles at the first stage of MA.

4. Conclusion

Common and distinctive features of mechanical alloying of Fe with sp-elements have been considered. The common regularities are following: The formation of a nanostructural state in α-Fe particles, sp-atoms penetration along the α-Fe grain boundaries, their segregation and the first Fe–M phase formation in the interfaces (boundary and close-to-boundary distorted zones) at the initial stage; the realization of any type of SSRs only on reaching the nanocrystalline state. The differences in the type of SSRs and their kinetics are conditioned by the ratio of the covalent radii, external shell electron configuration of sp-atom and sp-element concentration (x) in the initial mixture. In alloying α-Fe with sp-elements (Al, Si, Ge, Sn) having approximately equal and substantially larger atomic size the intermetallic compounds are formed in interfaces at the first stage. At the final stage supersaturated solid solution (SSS) is formed in the grain bulk if $x \leq 32$ at.% Si (Ge, Sn) and ≤ 60 at.% Al. In the Fe–Al (Si, Ge) systems the sp-element concentration in SSS becomes maximum simultaneously with the SSS formation, while in the Fe–Sn system SSS is saturated with Sn gradually. In α-Fe alloying with the C and B atoms of a small radius an amorphous phase (Am(Fe–M)) is formed in interfaces at the initial stage. The Am(Fe–B) formation is characterized by a substantially slower kinetics in comparison with that of the Am(Fe–C) one. If $x > 15$ at.% C(B) the second stage – the carbide and boride formation – takes place after amorphization. According to the results published by other authors and obtained in our studies we suggest the following scheme of mechanical alloying of Fe with sp-element (M), M = B, C, Al, Si, Ge and Sn (Figure 7).

Acknowledgments The authors express their gratitude to Prof. V.V. Boldyrev, Dr. T.F. Grigoryeva, Dr. G.N. Konygin, Dr. A.L. Ulyanov, Dr. V.A.Barinov and Dr. V.A.Zagainov for collaboration. This work has been supported by the Russian Fund for Basic Research (projects 97-03-33483 and 00-03-32555).

References

1. Yelsukov, E.P., Dorofeev, G.A., Barinov, V.A., Grigor'eva, T.F., Boldyrev, V.V.: Mater. Sci. Forum **269–272**, 151 (1998)
2. Dorofeev, G.A., Konygin, G.N., Yelsukov, E.P., Povstugar, I.V., Streletskii, A.N., Butyagin, P.Yu., Ulyanov, A.L., Voronina, E.V.: In: Miglierini, M., Petridis, D. (eds.) Mössbauer Spectroscopy in Materials Science. p. 151. Kluwer, the Netherlands (1999)
3. Dorofeev, G.A., Yelsukov, E.P., Ulyanov, A.L., Konygin, G.N.: Mater. Sci. Forum **343–346**, 585 (2000)
4. Dorofeev, G.A., Ulyanov, A.L., Konygin, G.N., Elsukov, E.P.: Phys. Met. Metallogr. **91**(1), 47 (2001)
5. Elsukov, E.P., Dorofeev, G.A., Konygin, G.N., Fomin, V.M., Zagainov, A.V.: Phys. Met. Metallogr. **93**(3), 93 (2002)
6. Yelsukov, E.P., Dorofeev, G.A.: Chem. Sustain. Dev. **1**, 243 (2002)

7. Elsukov, E.P., Dorofeev, G.A., Fomin, V.M., Konygin, G.N., Zagainov A.V., Maratkanova, A.N.: Phys. Met. Metallogr. **94**(4), 43 (2002)
8. Yelsukov, E.P., Dorofeev, G.A., Fomin, V.M.: J. Metastable Nanocryst. Mater. **15–16**, 445 (2003)
9. Elsukov, E.P., Dorofeev, G.A., Ulyanov, A.L., Nemtsova, O.M., Porsev, V.E.: Phys. Met. Metallogr. **95**(2), 60 (2003)
10. Elsukov, E.P., Dorofeev, G.A., Ulyanov, A.L., Zagainov, A.V.: Phys. Met. Metallogr. **95**(5), 486 (2003)
11. Yelsukov, E.P., Dorofeev, G.A., Boldyrev, V.V.: Doklady Chemistry **391**(4–6), 206 (2003)
12. Tanaka, T., Nasu, S., Ishihara, K.N., Shingu, P.H.: J. Less-Common Met. **171**, 237 (1991)
13. Suryanarayana, C.: Prog. Mater. Sci. **46**, 1 (2001)
14. Elsukov, E.P., Dorofeev, G.A., Ulyanov, A.L., Zagainov, A.V., Maratkanova, A.N.: Phys. Met. Metallogr. **91**(3), 46 (2001)
15. Trudeau, M.L., Schulz, R.: Mater. Sci. Eng. **A134**, 1361 (1991)
16. Horita, Z., Smith, D.J., Furakawa, M., Nemoto, M., Valiev, R.Z., Langdon, T.G.: Mater. Charact. **37**, 285 (1996)
17. Le Caër G., Matteazzi, P.: Hyperfine Interact. **66**, 309 (1991)
18. Wang, G.M., Calka, A., Campbell, S.J., Kaczmarek, W.A.: Mater. Sci. Forum **179–181**, 201(1995)
19. Campbell, S.J., Wang, G.M., Calka, A., Kaczmarek, W.A.: Mater. Sci. Eng. **A226–228**, 75 (1997)
20. Le Caër, G., Bauer-Grosse, E., Pianelly, A., Bouzy, E., Matteazzi, P.: J. Mater. Sci. **25**, 4726 (1990)
21. Tokumitzu, K.: Mater. Sci. Forum **235–238**, 127 (1997)
22. Tokumitzu, K., Umemoto, M.: Mater. Sci. Forum **360–362**, 183 (2001)
23. Nadutov, V.M., Garamus, V.M., Rawers, J.C.: Mater. Sci. Forum **343–346**, 721 (2000)
24. Bauer-Grosse, E., Le Caër, G.: Phys. Mag. **B56**, 485 (1987)
25. Okumura, H., Ishihara, K.N., Shingu, P.H., Park, H.S.: J. Mater. Sci. **27**, 153 (1993)
26. Calka, A., Radlinski, A.P., Shanks, R.: Mater. Sci. Eng. **A133**, 555 (1991)
27. Jing, J., Calka, A., Campbell, S.J.: J. Phys., Condens. Matter **3**, 7413 (1991)
28. Barinov, V.A., Tsurin, V.A., Elsukov, E.P., Ovechkin, L.V., Dorofeev, G.A., Ermakov, A.E.: Phys. Met. Metallogr. **74**(4), 412 (1992)
29. Balogh, J., Kemeny, T., Vincze, I., Bujdoso, L., Toth, L., Vincze, G.: J. Appl. Phys. **77**(10), 4997 (1995)
30. Passamani, E.C., Tagarro, J.R.B., Larika, C., Fernandes, A.A.R.: J. Phys., Condens. Matter **14**, 1975 (2002)
31. Chien, C.L., Musser, D., Gyorgy, E.M., Sherwood, R.C., Chen, H.S., Luborsky, F.E., WaLter, J.L.: Phys. Rev. **B20**, 283 (1979)
32. Barinov, V.A., Dorofeev, G.A., Ovechkin, L.V., Elsukov, E.P., Ermakov, A.E.: Phys. Met. Metallogr. **73**(1), 93 (1992)
33. Bansal, C., Gao, Z.Q., Hong, L.B., Fultz, B.: J. Appl. Phys. **76**, 5961 (1994)
34. Gaffet, E., Malhouroux, N., Abdellaoui, M.: J. Alloys Compd. **194**, 339 (1993)
35. Abdellaoui, M., Barradi, T., Gaffet, E.: J. Alloys Compd. **198**, 155 (1993)
36. Abdellaoui, M., Gaffet, E., Djega-Mariadassou, C.: Mater. Sci. Fofum **179–181**, 109 (1995)
37. Števulova, N., Buchal, A., Petrovic, P., Tkačova, K., Šepelak, V.: J. Magn. Magn. Mater. **203**, 190 (1999)
38. Cabrera, A.F., Sanchez, F.H., Mendoza-Zelis: Mater. Sci. Forum **312–314**, 85 (1999)
39. Sarkar, S., Bansal, C., Chatterjee, A.: Phys. Rev. **B62**, 3218 (2000)
40. Cabrera, A.F., Sanchez, F.H.: Phys. Rev. **B65**, 094202–1 (2002)
41. Nasu, S., Shingu, P.H., Ishihara, K.N., Fujita, F.E.: Hyperfine Interact. **55**, 1043 (1990)
42. Nasu, S., Imaoka, S., Morimoto, S., Tanimoto, H., Huang, B., Tanaka, T., Kujama, J., Ishihara, K.N., Shugu, P.H.: Mater. Sci. Forum **88–90**, 569 (1992)
43. Le Caër, G., Delcroix, P., Kientz, M.O., Malaman, B.: Mater. Sci. Forum **179–181**, 469 (1995)
44. Kientz, M.O., Le Caër, G., Delcroix, P., Fournes, L., Fultz, B., Matteazi, P., Malaman, B.: NanoStruct. Mater. **6**, 617 (1995)
45. Butyagin, P.Yu.: Chem. Rev. **B23**(Part 2), 89 (1998)
46. Schwarz, R.B.: Mater. Sci. Forum **269–272**, 663 (1998)
47. Shingu, P.H., Huang, B., Kuyama, J., Ishihara, K.N., Nasu, N.: In: Artz, E., Schultz, L. (eds.) New Materials by Mechanical Alloying Techniques, p. 319. DGM Informationgeselschaft, Oberursel (1989) p. 319
48. Wang, W.H., Xiao, K.Q., Dong, Y.D., He, Y.Z., Wang, G.M.: J. Non-Cryst. Solids **124**, 82 (1990)
49. Wang, G.M., Zhang, D.Y., Wang, W.H., Dong, Y.D.: J.Magn. Magn. Mater. **97**, 73 (1991)

50. Dong, Y.D., Wang, W.H., Liu, L., Xiao, K.Q., Tong, S.H., He, Y.Z.: Mater. Sci. Eng. **A134**, 867 (1991)
51. Guo, W., Martelli, S., Padella, F., Magini, M., Burgio, N., Paradiso, E., Franzoni, U.: Mater. Sci. Forum **89–90**, 139 (1992)
52. Bonetti, E., Scipione, G., Vadre, G., Cocco, G., Frattini, R., Macri, P.P.: J. Appl. Phys. **74**(3), 2058 (1993)
53. Fadeeva, V.F., Leonov, A.V., Khodina, L.M.: Mater. Sci. Forum **179–181**, 397 (1995)
54. Bonetti, E., Scipione, G., Frattini, R., Enzo, S., Schiffini, L.: J. Appl. Phys. **79**(10), 7537 (1996)
55. Enzo, S., Frattini, R., Gupta, R., Macri, P.P., Principi, G., Schiffini, L., Scipione, G.: Acta Mater. **43**, 3105 (1996)
56. Enzo, S., Mulas, G., Frattini, R.: Mater. Sci. Forum **269–272**, 385 (1998)
57. Enzo, S., Frattini, R., Mulas, G., Delogu, F.: Mater. Sci. Forum **269–272**, 391 (1998)
58. Wolski, K., Le Caër, G., Delcroix, P., Fillit, R., Thevenot, F., Le Coze, J.: Mater. Sci. Eng. **A207**, 97 (1996)
59. Jartych, E., Zurawicz, J.K., Oleszak, D., Sarzynski, J., Budzynski, M.: Hyperfine Interact. **99**, 389 (1996)
60. Oleszak, D., Shingu, P.H., Matyja, H.: In: Duhaj, P., Mrafko, P., Svec, P. (eds.) Rapidly Quenched and Metastable Materials, Supplement, p. 18. Elsevier, Notherlands (1997)
61. Oleszak, D., Shingu, P.H.: Mater. Sci. Forum **235–238**, 91 (1997)
62. Oleszak, D., Pekala, M., Jartych, E., Zurawicz, J.K.: Mater. Sci. Forum **269–272**, 643 (1998)
63. Jartych, E., Zurawicz, J.K., Oleszak, D., Pekala, M.: J. Phys., Condens. Matter **10**, 4929 (1998)
64. Eelman, D.E., Dahn, J.R., Mackkay, G.R., Dunlap, R.A.: J. Alloys Compd. **266**, 234 (1998)
65. Hashii, M.: Mater. Sci. Forum **312–314**, 139 (1998)
66. Hashii, M., Tokumitsu, K.: Mater. Sci. Forum **312–314**, 399 (1999)
67. Gonzales, G., D'Angelo, L., Ochoa, J., D'Onofrio, L.: Mater. Sci. Forum **360–362**, 349 (2001)
68. Yelsukov, E.P., Voronina, E.V., Barinov, V.A.: J. Magn. Magn. Mater. **115**, 271 (1992)
69. Perez Alcazar G.A., Galvao da Silva, B.: J. Phys. F. Met. Phys. **17**, 2323 (1987)

Hyperfine Interact (2005) 164: 67–72
DOI 10.1007/s10751-006-9234-4

Improvements in Depth Selective Electron Mössbauer Spectroscopy

**N. U. Aldiyarov · K. K. Kadyrzhanov ·
E. M. Yakushev · V. S. Zhdanov**

Abstract The conditions for achievement of high resolving power of depth selective conversion electron Mössbauer spectroscopy method at a combined installation 'electron spectrometer–nuclear gamma-resonance spectrometer' have been obtained. There has been made a considerable step in development of the method at its realization at a magnetic sector electron spectrometer with double focusing, equipped with electron source (a sample under investigation) of large-area and position-sensitive detector. The paper presents a prospective symmetrical version of a magnetic sector electron spectrometer that allows realizing more completely possibilities of the method. It is noted that the proposed method is particularly valuable for investigations of nanosystems, nanostrutures that contain Mössbauer nuclei.

Key words Depth Selective Conversion Electron Mössbauer Spectroscopy (DS CEMS) · Depth Selective Electron Mössbauer Spectroscopy (DS EMS) · electron spectrometer · nanotechnology.

1. Introduction

Broad application of Mössbauer spectroscopy in various investigations including study of material properties at microscopic level stipulates continuous improvement of its instrumentation. During the last years, in particular, there has been actively developed a method of non-destructive depth selective conversion electron Mössbauer spectroscopy (DS CEMS), which allows to obtain Mössbauer information from subsurface nanolayers of samples with high depth resolution. Precise investigations by DS CEMS method at a combined installation 'electron spectrometer–nuclear

N. U. Aldiyarov · K. K. Kadyrzhanov · E. M. Yakushev · V. S. Zhdanov (✉)
Institute of Nuclear Physics NNC RK, 1 Ibragimov Str., Almaty 480082, Kazakhstan
e-mail: zhvs@inp.kz

gamma-resonance spectrometer' are of special interest. Quality of such investigations depends first of all upon the possibilities of the electron spectrometer used.

2. Conditions for achievement of high resolving power of depth selective conversion electron Mössbauer spectroscopy

A natural physical limit of depth resolution for the DS CEMS method is stipulated by mean free path of inelastically scattered electrons. It is known [1] that free path for electrons reduces from ~5 up to ~0.5 nm within the energy range from units up to 10 eV, then from 10 up to 500 eV there is a wide minimum of ~0.5 nm and above 500 eV it increases up to ~5 nm. Experimentally, due to some reasons, depth resolution does not reach this limit; therefore it is possible, having minimized influence of these reasons, to improve depth resolving power. Let us consider in more details the electron spectrometer sensitive conditions for achievement of depth resolution of the method close to its physical limit.

Available electron spectrometers of combined installation are quite different by the escape angle of electrons with respect to normal to the source (sample) surface (see, for instance [2–4]). So, there were performed model calculations using Monte-Carlo method of angular and energy distributions for electrons from various depths. One should keep in mind that depth resolution of the DS CEMS method is directly related to abruptness of changes of contributions from different depth layers into the registered intensity at electron line. Not being sunk in calculation details, we mention here two important results: At escape of monoenergy electrons from some depth that exceeds free path, width of energy distribution of scattered electrons is increased with rise of the escape angle relative to the surface normal, and position of energy distribution maximum shifts to low energies. We would like to note that the widening of energy distributions at increase of angle of escape exceeds bias of their maximums. This results in lower change in contributions into the measured intensity from different depth layers and, correspondingly, in lower depth resolution of the method. Thus, selection of an angle of electrons' escape from a sample is of essential significance at realization of DS CEMS experiments; the selection practically means selection of the electron spectrometer type for the combined installation. To improve depth resolving power one needs electron spectrometers with energy resolution better than 1% that enable operation at escape angles relative to a normal to a surface of an electron source close to zero [5]. Selection of such preferred angle was verified experimentally. As well, in some cases there is a necessity to operate at different angles.

Resolving power of DS CEMS method is immediately determined by a statistically significant number of layers the investigated thickness is split up for, that is required for mathematical processing of Mössbauer spectra. Naturally, it directly depends on statistics accumulated on the spectra. The role and magnitude of the ratio effect/background are essential as well. This further narrows down a set of electron spectrometers most eligible for DS CEMS method, i.e. those, available for further improvement of depth resolution.

So, to reach high depth resolving power in DS CEMS method one needs a highly effective 'precise' electron spectrometer with excellent ratio effect/background that operates mainly at close to zero escape angles (with respect to the normal to a sample surface).

 Springer

3. Improvements in depth selective electron Mössbauer spectroscopy

In the last decades the improvement of efficiency of 'precise' electron spectrometers was mainly realized in two basis directions. One of them is connected with searching of new configurations for electric and magnetic fields or their combination to achieve acceptable focusing properties. Progress in this field seems to be quite moderate. We note here that increase in effectiveness due to considerable increase of solid angle is unreasonable since it results in lower depth resolution of the DS CEMS method. In the other direction there are used the possibilities of currently available spectrometer types, in particular, availability at some of them of focal plains both in the region of a source, and in the region of a detector. Electron spectrometers with a transverse heterogeneous magnetic field of $1/\sqrt{\rho}$ type that provides double focusing can serve as an example. Application of non-equipotential large-area sources or position registration allows speeding up for many times accumulation of spectral information without noticeable loss in resolution. At that the device efficiency can be increased in the first case by one order of magnitude, in the second case – by two orders of magnitude. As a whole, this for sure exceeds the restricted solid angle of such instruments in comparison with some other types of spectrometers.

In [3] we have used a combined installation for depth selective conversion electron Mössbauer spectroscopy a sector electron spectrometer with transverse heterogeneous magnetic field that provided double focusing.[1] This spectrometer meets the presented above conditions. Sakai-type electron spectrometer with $1/\sqrt{\rho}$ magnetic field and extensive focal plans at source and detector [7] was chosen as a prototype for the spectrometer [3]. A schematic drawing of our electron spectrometer is presented at Figure 1. Main characteristics of its improved variant are as follows: Radius of equilibrium orbit $\rho_o = 120$ mm; instrumental energy resolution 0.4% at solid angle 0.6% of 4π steradian; electron source (sample) dimensions 0.7×15 mm^2. The spectrometer is equipped with a non-equipotential multi-strip electron source 3 of Bergkvist' type [8] similar to [9] and a position-sensitive detector 4 on the basis of microchannel plate chevron [10]. Total area of the multi-strip electron source (a sample) achieves 150 mm^2. Potentials to adjust strip images in the spectrometer focal plane towards detector are delivered to the strips of this source. The position-sensitive detector has a rectangular inlet window of 40 mm in width and 17 mm in height that allows registration of the part of electron spectrum in relative energy interval $\sim$10%. It is quite enough for implementation of the DS CEMS analysis by several Mössbauer spectra measured 'simultaneously'. Angles of electrons escape relative to the normal to a sample surface in electron spectrometer are close to zero. There is a possibility to work at other angles as well.

Beyond conversion electrons, the proposed electron spectrometer [3] also allows working with low-energy electrons (up to 500 eV) including Auger, Coster–Kroning electrons and electrons of true secondary emission. Thickness of about several nm with depth resolution of about 10th fractions of an nm is studied in the last case with rather high efficiency of the measurements.

Simultaneous utilization of a multi-strip electron source (a sample) and a position-sensitive detector in the electron spectrometer faces some known difficulties. The

---

[1] For the first time sector magnetic spectrometer with $1/\rho$ field was used as CEMS-spectrometer in 1961 [6].

Figure 1 Schematic drawing of
magnetic sector electron
spectrometer ($\rho_0 = 120$ mm)
of combined installation for
DS CEMS. 1 – vacuum
camera, 2 – sector poles,
3 – electron source (sample),
4 – position–sensitive detector,
5 – beryllium window,
6 – Mössbauer effect driver,
7 – Mössbauer source,
8 – lead protection.

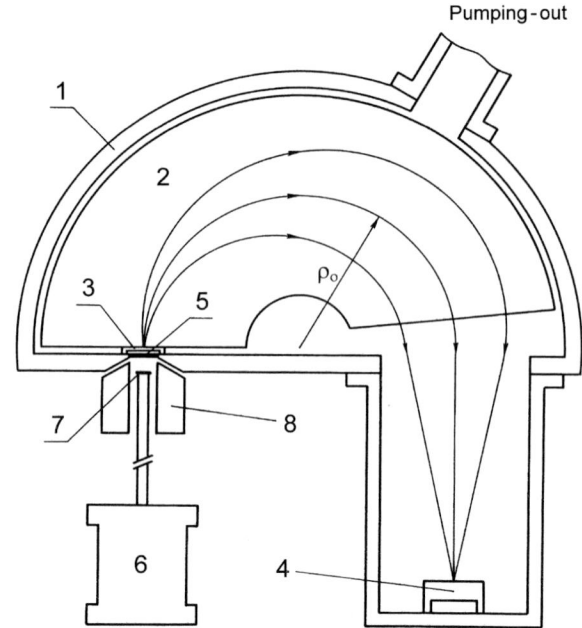

multi-strip source provides an excellent focusing in central part of the detector. Moving from the detector center to its edges (along its width), focusing becomes worse due to increase distinction of potentials at source strips from the optimal values [8]. This results in some worsening of depth resolution. From the other side, utilization of a multi-strip source assures considerable increase in statistics for Mössbauer spectra, i.e. improves achieved depth resolution. So, using multi-strip sources of various widths there was experimentally found[2] optimal width for the multi-strip source that comprised 12 mm to provide the highest depth resolution [11].

Let us compare the improved version of this magnetic sector electron spectrometer [3] with a magnetic high luminosity multi-gap spectrometer of 'orange' type [2]. The spectrometer [2] considerably outperforms [3] in solid angle. But, due to utilization in [3] of a large-area non-equipotential electron source (a sample) and high resolution, both instruments provide similar experimental possibilities when consider resolution-luminosity (see Figure 3 in [2].) In general, magnetic spectrometers [2] and [3] compared to other specialized but electrostatic spectrometers are in more favorable position. This is particularly so when consider utilization of position detection in both magnetic spectrometers. One should note that much better ratio of a sample size to dispersant element dimension for the electron spectrometer [3] and comparative simplicity of its design would make it possible in the future to use it as a basis for development of a portable magnetic spectrometer.

Although, the magnetic sector electron spectrometer [3] with a sample at the boundary of the magnetic gap corresponds to maximum solid angle, but thoroughly

---

[2] Here the distance between the exciting Mössbauer source ($\sim$100 mCi of 57Co Pd) and a multi-strip electron source (a sample) exceeded the size of the last for $\sim$5 times.

 Springer

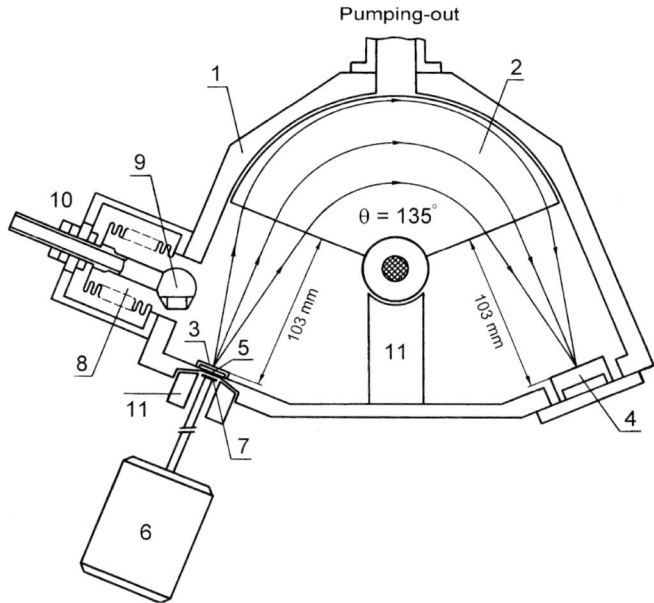

Figure 2 Schematic drawing of prospective symmetrical variant of magnetic sector electron spectrometer ($\rho_o = 80$ mm) of combined installation for DS EMS. 1 – vacuum camera, 2 – symmetrical sector poles, 3 – electron source (sample), 4 – position-sensitive multi-detector, 5 – beryllium window, 6 – Mössbauer effect driver, 7 – Mössbauer source, 8 – additional camera, 9 – ion gun, 10 – operating mechanism of ion gun, 11 – lead protection.

calculated prospective symmetrical variant of the spectrometer (see Figure 2) at small loss in solid angle allows, due to the arrangement of a sample outside of the interpolar gap, to realize more completely the possibilities of the method [12]. First of all, this refers to organization of experiments in any geometry; secondly, to carrying out of investigations in wide temperature range including low temperatures; thirdly, to preparation and modification of investigated nanolayers; and finally, to the possibility to detect gamma-quanta passing through a sample. A simplified position-sensitive detector [13] with multi-collector in form of strips that repeat the image form in focal plane of the spectrometer is used in the prospective symmetrical variant of the magnetic sector electron spectrometer. Maximal count rate of the detector in this case is increased for two orders of magnitude. An additional dispersant element is mounted in front of the detector in order to pass in detail the part of electron line corresponding to one collector [13].

The main aspects of application of internal conversion electrons, Auger electrons and secondary electrons that accompany decay of Mössbauer levels of different nuclei were considered in details for the method of depth selective electron Mössbauer spectroscopy (DS EMS) [14]. There were recommended optimal variants for use of either type of electron irradiation and their combinations for the majority of the tasks. A set of these recommendations has been verified at the combined installment.

Prospects of application of the depth selective electron Mössbauer spectroscopy method depend on the possibility to obtain the experimental information from nanolayer to nanolayer within the limits of the absorption thickness of corresponding

electrons that comprises from ten to hundreds of nanometers. The method is obviously promising for low-temperature investigations of nanosystems and nanostructures that contain Mössbauer nuclei and became highly topical owing to the development of nanotechnology [15]. There are in progress tests and developments of the procedure for realization of such investigations at a combined installation 'electron spectrometer–nuclear gamma-resonance spectrometer' using model nanolayer samples.

References

1. Gallon, T.E.: In: Fiermans, L., Vennik, J., Dekeyser, W. (eds.) Electron and Ion Spectroscopy of Solids. Plenum, New York and London (1978)
2. Stahl, B., Kankeleit, E.: Nucl. Instr. Meth. Phys. Res. B. **122**, 149 (1997)
3. Ryzhykh, V.Yu., Babenkov, M.I., Bobykin, B.V., Zhdanov, V.S., Zhetbaev, A.K.: Nucl. Instr. Meth. Phys. Res. B. **47**, 470 (1990); ICAME'1989, Abstracts, Budapest, Hungary, 1989, p.17.15
4. Toriyama, T., Asano, K., Saneyoshi, K., Hisatake, K.: Nucl. Instr. Meth. Phys. Res. B. **4**, 170 (1984)
5. Babenkov, M.I., Zhdanov, V.S., Ryzhykh, V.Yu., Chubisov, M.A.: LI International Conference on Nuclear Spectroscopy and Nuclear Structure, Abstracts, Sarov, Russia, p.259, 2001
6. Kankeleit, E.: Z. Physic. **164**, 442 (1961)
7. Sakai, M., Ikegami, H., Yamazaki, T.: Nucl. Instr. Meth. **9**, 154 (1960)
8. Bergkvist, K.E.: Ark. Fys. **27**, 383 (1964)
9. Babenkov, M.I., Bobykin, B.V., Zhdanov, V.S., Petukhov, V.K.: J. Phys. B: At. Mol. Phys. **15**, 35 (1982)
10. Babenkov, M.I., Zhdanov, V.S., Starodubov, S.A.: Nucl. Instr. Meth. Phys. Res. A. **252**, 83 (1986)
11. Aldijarov, N.U., Zhdanov, V.S., Kadyrzhanov, K.K., Yakushev, E.M.: ICMSA'2002, Abstracts, St.-Petersburg, Russia, p.207, 2002
12. Aldijarov, N.U., Zhdanov, V.S., Kadyrzhanov, K.K., Yakushev, E.M.: ICMSA'2002, Abstracts, St.-Petersburg, Russia, p.206, 2002
13. Aldijarov, N.U., Zhdanov, V.S., Kadyrzhanov, K.K., Yakushev, E.M.: II Eurasian Conference on Nuclear Science and Its Application, Presentations, vol.2, Almaty, Kazakhstan, p.70, 2003
14. Babenkov, M.I., Zhdanov, V.S., Zhetbaev, A.K., Ryzhykh, V.Yu.: Preprint 1–93, INP NNC, Almaty, Kazakhstan, 1993
15. Aldiyarov, N.U., Kadyrzhanov, K.K., Ryzhykh, V.Yu., Zhdanov, V.S.: ICAME'2003, Abstracts, Muscat, Oman, p.T9/6, 2003

Hyperfine Interact (2005) 164: 73–85
DOI 10.1007/s10751-006-9235-3

Thermal Induced Processes in Laminar System of Stainless Steel – Beryllium

A. K. Zhubaev · K. K. Kadyrzhanov · V. S. Rusakov ·
T. E. Turkebaev

Abstract The paper reports on investigation of the laminar system 'stainless steel 12Cr18Ni10Ti – Be' at thermal treatment. There have been determined sequences of phase transformations along with relative amount of iron-containing phases in the samples subjected to thermal beryllization. It has been revealed that thermal beryllization of stainless steel thin foils results in $\gamma \rightarrow \alpha$ transformation and formation of the beryllides NiBe and FeBe$_2$. It has also been revealed that direct $\gamma \rightarrow \alpha$- and reverse $\alpha \rightarrow \gamma$-transformations are accompanied by, correspondingly, formation and decomposition of the beryllide NiBe. It is shown that distribution of the formed phases within sample bulk is defined by local concentration of beryllium. Based on obtained experimental data there is proposed a physical model of phase transformations in stainless steel at thermal beryllization.

Key words stainless steel · plastic deformation · thermal annealing · beryllides · beryllization · phase transformations.

1. Introduction

Development of technology rises up creation of materials that operate at extreme conditions of high temperatures, mechanical loads, aggressive media and external irradiation. At the same time different requirements are set for the properties of material surface and bulk what leads to development of new approaches to creation of alloys in order to assure reliable multi-function performance of such newly designed materials.

A. K. Zhubaev (✉) · K. K. Kadyrzhanov · T. E. Turkebaev
Institute of Nuclear Physics, National Nuclear Center, Almaty 050032, Kazakhstan
e-mail: zhubaev@inp.kz

V. S. Rusakov
Physics Department, Moscow State University, Moscow 119899, Russia

Table I Element composition of the steel 12Cr18Ni10Ti – Main components

Element	C	Cr	Ni	Mn	Si	Ti	Fe
Content (wt.%)	0.1	17.3	9.7	0.62	~0.54	~0.51	Basis

For instance, the system [beryllium-alloy based on copper-construction steel] is considered for application in nuclear power production. Beryllium containing alloys demonstrate unique physical and mechanical properties [1]. In this connection investigation of beryllium interaction with stainless steel is topical.

Several works are devoted to investigations of beryllium interaction with stainless steel (see, for instance, [2–7]) performed, mainly, in 1960–1970s and generalized in the monographs [8–10]. In all experiments thickness of investigated samples exceeded significantly thickness of formed diffusion layers. In [2–5, 7] there were measured diffusion layer thicknesses a function of time in wide temperature range of $500 \div 1,000\,°C$; at that duration of thermal annealing was up to several thousands hours. For determination of diffusion layer thickness was used, as a rule, consequent electrochemical or mechanical treatment that could influence the result of thermal beryllization. In a diffusion layer were revealed formed beryllium alloy in steel [5–7], beryllides of iron [5, 7] and nickel [8], nickel segregations enriched with beryllium [6].

2. Experimental section

In order to decrease relaxation time of nonequilibrium thermally induced processes of diffusion and phase transformations as well as for effective application of Mössbauer spectroscopy and X-ray phase analysis, we used thin foils of stainless austenitic steel 12Cr18Ni10Ti as a substrate; see its element composition in Table I.

Mössbauer spectra were taken by registration of γ-quanta in absorption geometry (MS). In this case Mössbauer spectroscopy provides information on average phase state of the whole sample. We used in the investigations a ^{57}Co source in the Cr matrix with activity of about 30 mCi. MS- and CEMS-spectra were obtained at room temperature at the Mössbauer spectrometer CM2201 operating at constant accelerations and sawtooth time dependence of Doppler source velocity with respect to the absorber. Scintillation counter with NaJ(Tl) crystal was used registration of 14.4 keV γ-quanta. Spectrometer calibration was performed using a precise standard α-Fe enriched with ^{57}Fe up to 89%.

Processing and analysis of experimental Mössbauer spectra were performed using the software package MSTools [11]. Model decoding of MS-spectra for standard samples was realized using the code SPECTR [12]. Processing of Mössbauer spectra from investigated systems was made by reconstruction of distribution functions for the partial spectra hyperfine parameters [13]. The code DISTRI was used for processing of the spectra [11].

X-ray phase analysis of the samples was performed at automatic diffracrometer DRON-2 with β-filter and Cu–$_\alpha$ radiation. The measurements were performed in the Bragg–Brentano geometry from both sample sides. In this case were studied phase states of near-surface layers ($3 \div 4\,\mu$m). For identification of crystalline phases was used a card index of powder radiographs ASTM.

 Springer

When traditional technologies for foil production are used scientific approach is needed. In particular, one should know how plastic deformation influences structure of original (before rolling) steel and regimes of thermal annealing for initial structure recovery.

3. Results and discussion

3.1. Plastic deformation

Investigation samples were prepared from 135 μm-thick plain steel 12Cr18Ni10Ti by cold rolling at the forging rolls of hardened carbon steel to the thickness $\sim$30 μm. Deformation rate comprised $\sim$53%. Thermal treatment was performed in a vacuum furnace at residual pressure 5×10^{-5} mmHg. The temperature was controlled by a tungsten-rhenium thermal couple as precise as $\pm$5 °C. There were performed 1-h isochronous annealings of deformed samples at temperatures $200 \div 900$ °C.

Experimental MS-spectra for rolled up and annealed stainless steel foils were presented by a superposition of two partial spectra – Paramagnetic Mössbauer line and Zeeman sextet (see Figure 1). At that, higher the annealing temperature, lower relative intensity of the partial spectrum for Zeeman sextet that corresponds to martensite (α-phase and higher the intensity of paramagnetic Mössbauer line that corresponds to austenite (γ-phase).

Due to considerable widening of resonance lines and in order to reveal slight changes in spectra, processing and analysis of experimental spectra were performed by reconstruction of distribution functions for hyperfine parameters of partial spectra (see [11]). For each of the partial spectra there was chosen a hyperfine parameter distribution function for the parameter most sensitive to local heterogeneity of ^{57}Fe nuclei. For paramagnetic partial spectrum that was distribution function of quadruple shift $p(\varepsilon)$, and for the Zeeman sextet – Distribution function of hyperfine magnetic fields $p(H_n)$.

According to [14] quadruple shift of Mössbauer spectrum components for austenitic steel comprises $\varepsilon \cong 0.08$ mm/s. In our case reconstructed distribution functions for quadruple shift $p(\varepsilon)$ are single-mode functions with a maximum lying right in vicinity of this value. Therefore, paramagnetic partial spectrum is associated only with Fe atoms in austenitic state γ of investigated stainless steel.

It has been revealed that at higher annealing temperatures the intensity of partial spectrum for ferromagnetic phase (martensite) decreases and, upon annealing at $\sim$650 °C, there is the only paramagnetic line of austenite in the spectrum.

Upon each annealing stage we run X-ray phase analysis of deformed and annealed samples of stainless steel. X-ray diffarctogram of the rolled up steel (deformation rate 53%) demonstrated reflexes of martensite (α-phases; see Figure 1). Thermal annealing results in formation of wider reflexes for γ-phase. Further annealing temperature increase eliminates α-phase reflexes. After annealing at 650 °C the diffractogram of a sample contains austenite reflexes only.

Upon processing of obtained MS-spectra there were obtained annealing temperature dependencies for relative intensity of α-phase partial spectrum in the bulk of the investigated samples (Figure 2).

Some parts of X-ray diifractograms ($40° \leq 2\theta \leq 47°$ and $70° \leq 2\theta \leq 85°$) were processed using the codes PREPSPEC and SPECTR [11] at description of X-ray

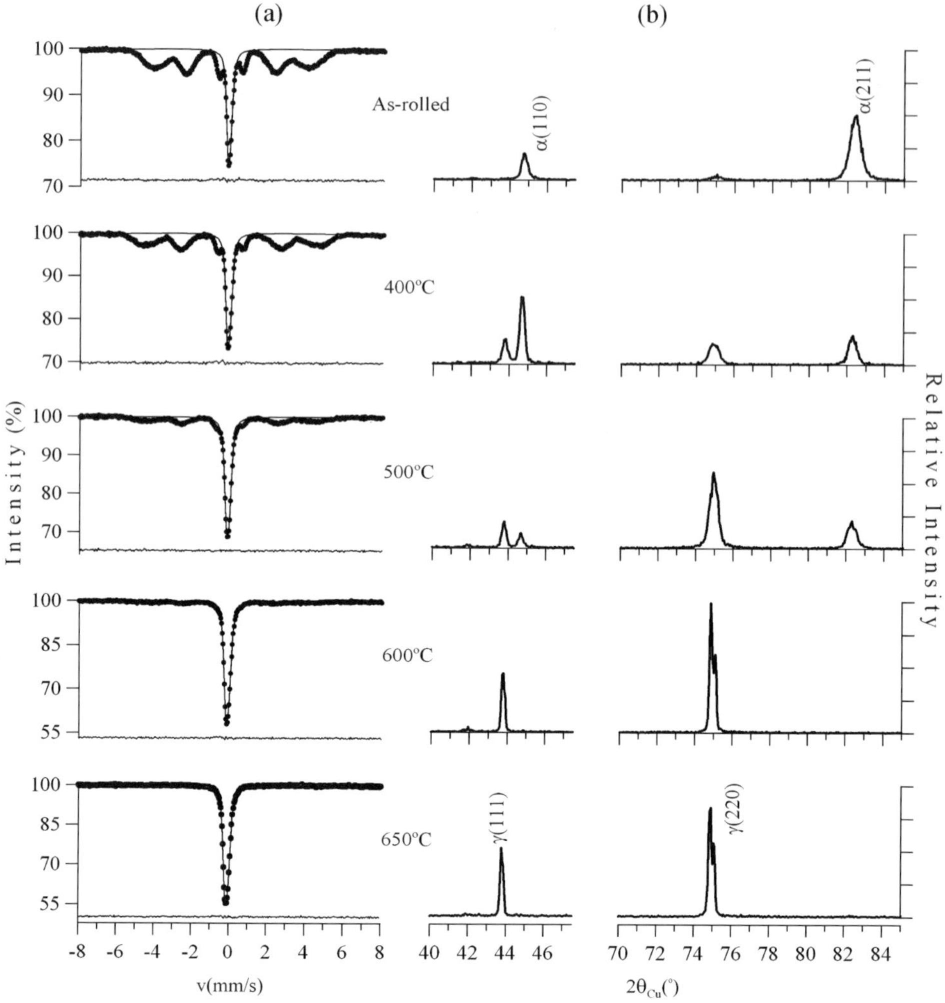

Figure 1 MS-spectra and X-ray difractograms for stainless steel 12Cr18Ni10Ti upon plastic deformation with 53% squeezing and 1 h annealing at various temperatures.

reflexes with the line Pseudo–Voigt. In order to assess relative martensite content in the sample, there were obtained relative intensities of the reflexes $\alpha 110/\gamma 111$ and $\alpha 211/\gamma 220$ as functions of thermal annealing temperatures (Figure 2).

Based on consideration of mutual arrangement of the curves at the figure one can see that reverse $\alpha \rightarrow \gamma$-transformation at annealing starts at the surface and transformation rate at surface is higher. There have been revealed two characteristic temperature intervals of reverse transformations at tempering of the deformed steel. At temperatures below $\approx 400\,^{\circ}$C there is some insignificant $\alpha \rightarrow \gamma$-transformation due to decrease of local concentration of chromium atoms (to $\sim 8\%$) while they diffuse from α-phase inclusions into defective structure regions. In the interval $400 \div 600$ $^{\circ}$C intensive $\alpha \rightarrow \gamma$-transformation takes place induced by thermal motion of atoms. Above 600 $^{\circ}$C complete recovery of austenite is observed.

 Springer

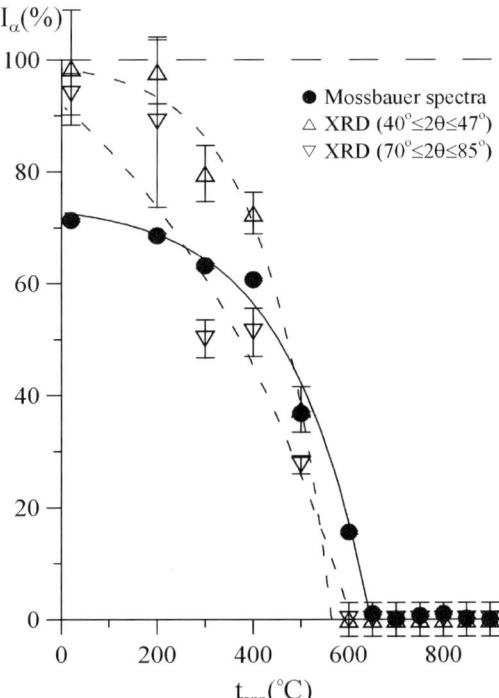

Figure 2 Relative intensities of Mössbauer partial spectrum of the martensite and diffraction lines $\alpha_{110}/\gamma_{111}$ and $\alpha_{211}/\gamma_{220}$ for a deformed sample with 60% squeezing *versus* annealing temperature.

3.2. Beryllization

This part discusses main results of Mössbauer spectroscopy and X-ray phase analysis phase transformation investigations at thermal annealings of stainless thin foils coated with beryllium.

Steel foil substrates were coated with beryllium of d_{Be} (1 and 2 μm) by magnetron sputtering at Argamak installment. The substrates were fixed at a massive holder kept (as well as the substrates) during sputtering at temperatures not exceeding 150 °C what made diffusion during sputtering insignificant. The substrate was subjected to etching with argon ions prior sputtering to assure better adhesion. Ion etching and beryllium sputtering were realized in a single vacuum cycle. Control over sputtered layer thickness was performed by weight method.

Thin foils coated with beryllium were subjected to successive isochronous annealings. One-hour isochronous annealings were performed in the temperature range 400 ÷ 800 °C with gradual increase for 50 °C. Thermal annealings were performed in a muffle furnace at pressure 6×10^{-6} mmHg. The given temperature was achieved for about 30 min with the same for cooling down of the samples together with the furnace.

In general, experimental spectrum is a superposition of two well-resolved partial spectra with different relative intensities – A paramagnetic Mössbauer line and a Zeeman sextet with noticeably widened components (see Figure 3). Therefore, processing of all the spectra obtained from the investigated beryllium-coated foils was done with reconstruction of distribution functions for partial spectra hyperfine parameters – A distribution function $p(\varepsilon)$ of quadruple shift ε and a distribution

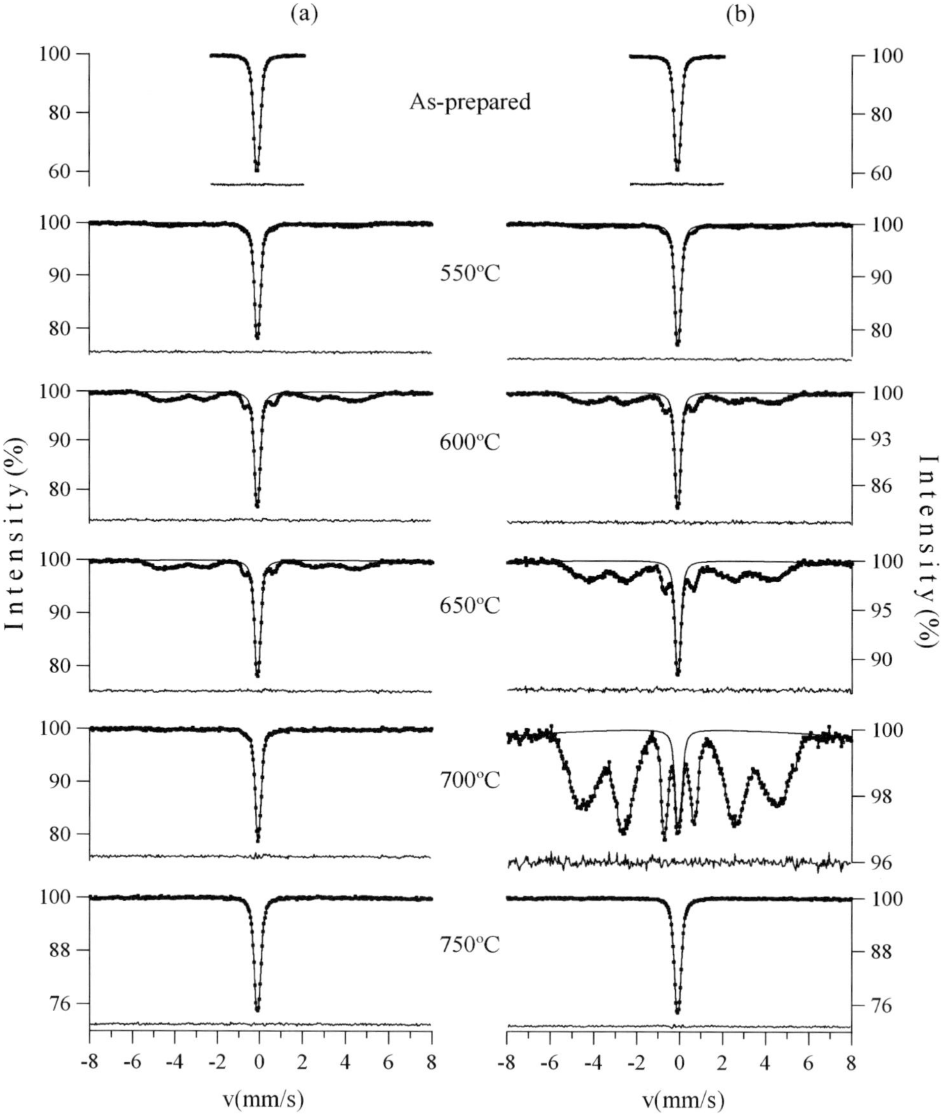

(a)   (b)

Intensity (%)

As-prepared

550°C

600°C

650°C

700°C

750°C

v(mm/s)   v(mm/s)

Intensity (%)

Figure 3 MS-spectra for stainless steel samples of thicknesses $d_{Be} = 1$ and 2 μm upon consequent isochronous 1-h annealings.

function $p(H_n)$ of effective magnetic field H_n. At that we searched for possible linear correlations between the hyperfine parameters for each of the partial spectra [13].

Steel foils with $d_{Be} = 1$ and 2 μm coatings were subjected to successive isochronous annealings within the temperature range $400 \div 800$ °C. At the initial stage there was observed $\gamma \rightarrow \alpha$-transformation in both samples started at 550 °C. At that in a sample with more thick coating relative intensity of partial spectrum of the α-phase (Zeeman sextet with widened lines) is considerably higher and reaches its maximal value ($\sim$90% *versus* $\sim$50% for a sample coated with $d_{Be} = 1$ μm) at higher temperature (see Figure 3). For the investigated foils, further annealing temperature t_{ann} increase

 Springer

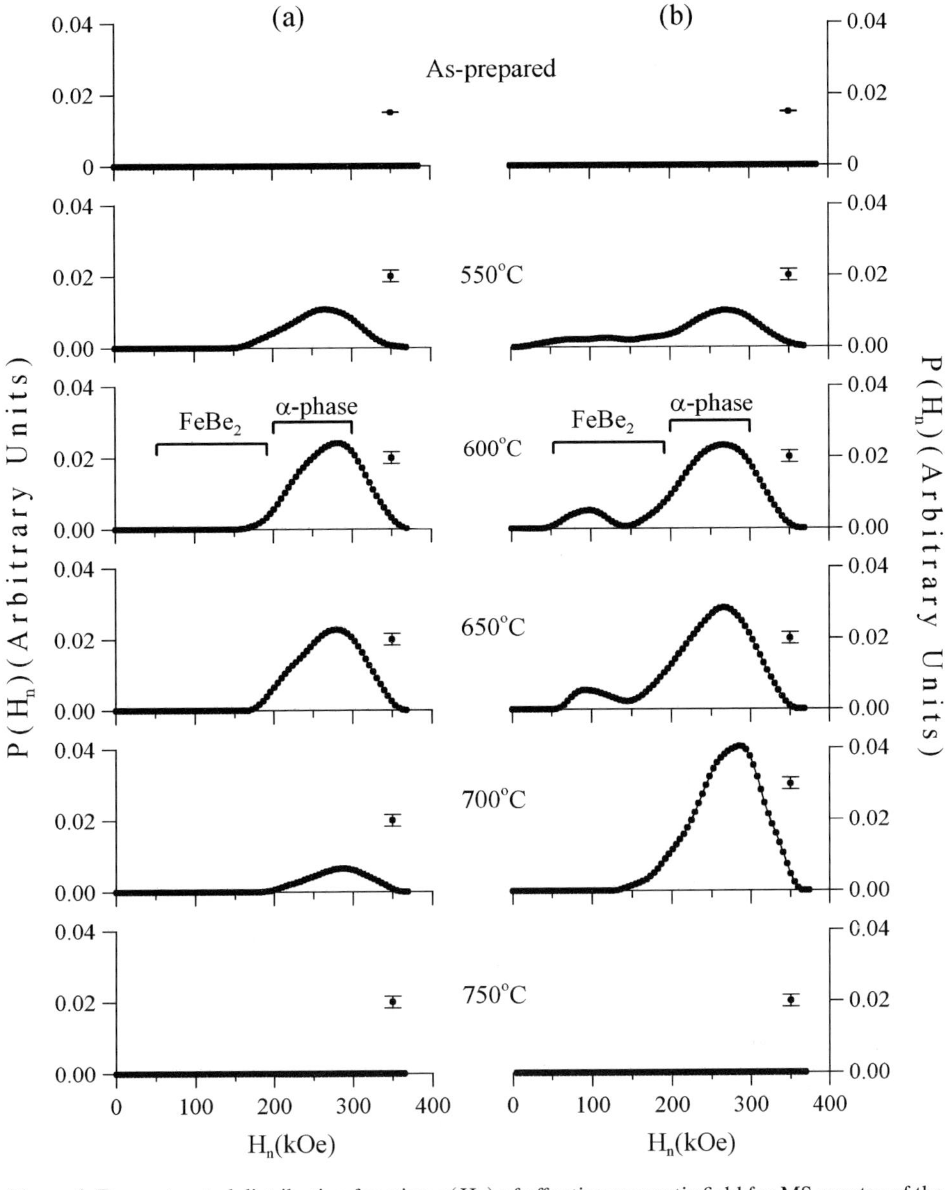

(a)   (b)

As-prepared

550°C

FeBe₂   α-phase   600°C   FeBe₂   α-phase

650°C

700°C

750°C

P(Hₙ)(Arbitrary Units)

H$_n$(kOe)   H$_n$(kOe)

Figure 4 Reconstructed distribution function $p(H_n)$ of effective magnetic field for MS-spectra of the samples with $d_{Be} = 1$ and $2\ \mu$m upon consequent isochronous 1-h annealings.

results in total reverse $\alpha \rightarrow \gamma$-transformation and MS spectra became represented by paramagnetic spectra that corresponded to austenite.

Upon reconstruction of distribution functions for partial Mössbauer spectra distribution functions, it was revealed (Figure 4), that distribution function of effective magnetic field $p(H_n)$ for a sample with $d_{Be} = 1\ \mu$m is of single-mode form with one local maximum at $H_n \sim 260$ kOe, and for a sample coated with $d_{Be} = 2\ \mu$m – Two-mode form with two local maxima at $H_n \sim 100$ and ~ 260 kOe.

 Springer

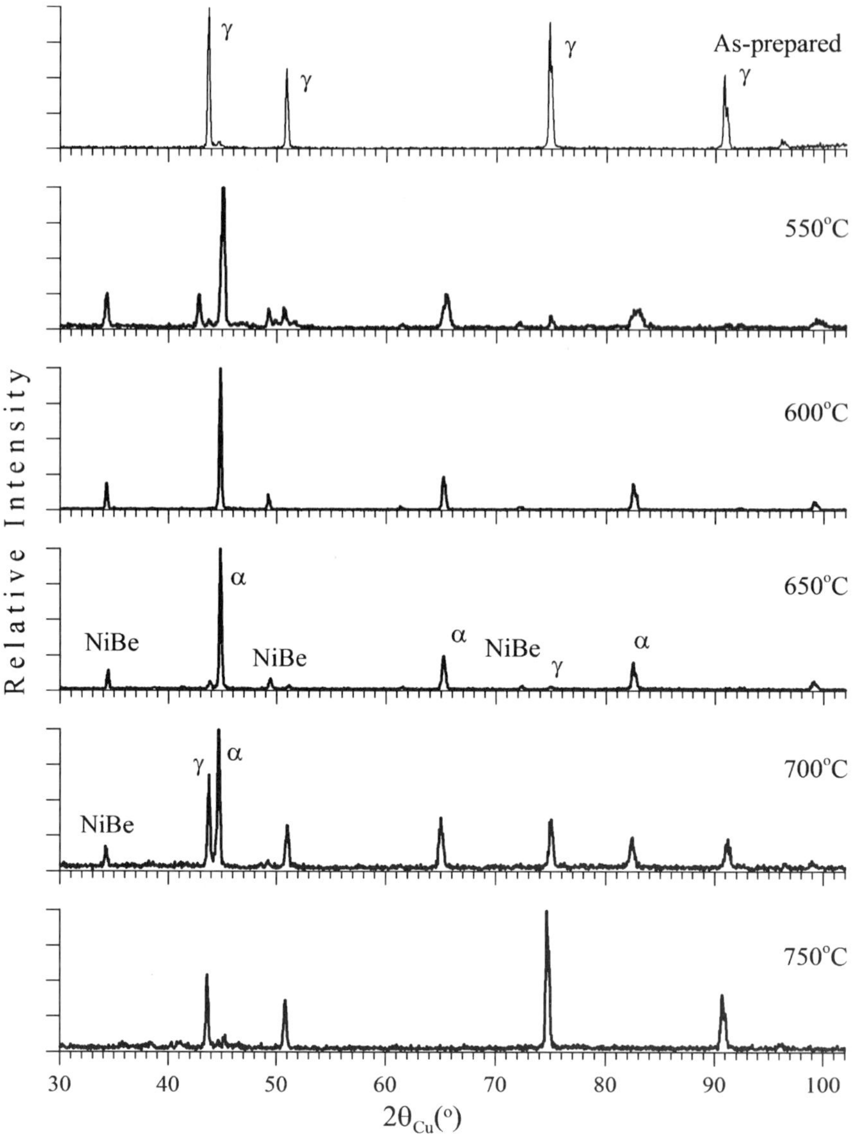

Figure 5 X-ray difractograms of the coating-facing side of the stainless steel samples of $d_{Be} = 1$ and 2 μm upon consequent isochronous 2-h annealings.

They can be uniquely identified using available literature data. The local maximum at the field range H_n 53 ÷ 192 kOe [15] can be associated with Fe atoms that belong to iron beryllide $FeBe_2$, and the local maximum at the energy range 200 ÷ 300 kOe [16] – With Fe atoms that belong to the steel in the martensite state (α).

In other words, if in the first case there is observed formation of α-phase only, in the second case there is also formation of the beryllide $FeBe_2$. So, one can conclude that introduction of beryllium into austenitic stainless steel, where all Fe atoms are

Springer

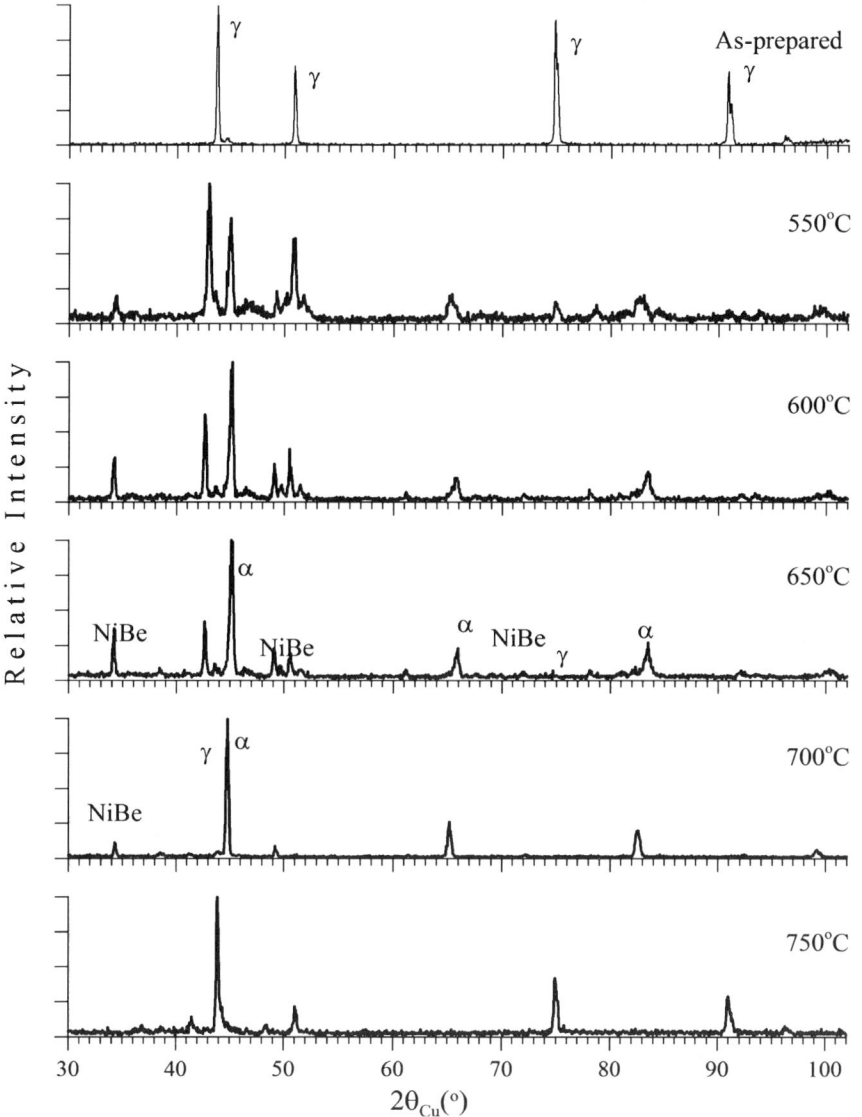

Figure 6 X-ray diffractograms of the backing side of the stainless steel samples of $d_{Be} = 1$ and $2\ \mu m$ upon consequent isochronous 2-h annealings.

in γ-phase, results in formation of α-phase, and at sufficient beryllium concentration there is also formation of the beryllide $FeBe_2$.

Analysis of reconstructed distribution functions for quadruple shift $p(\varepsilon)$ and of effective magnetic field $p(H_n)$ for the range of hyperfine parameters characteristic for various iron-containing phases performed using the code DISTRI [11, 13] enabled us to obtain relative intensities for partial Mössbauer spectra I_M as a function of isochronous annealing temperature (see Figure 5). As one can see, thermal annealing of a stainless steel thin foil coated with $d_{Be} = 1\ \mu m$ at 550 °C results in formation of

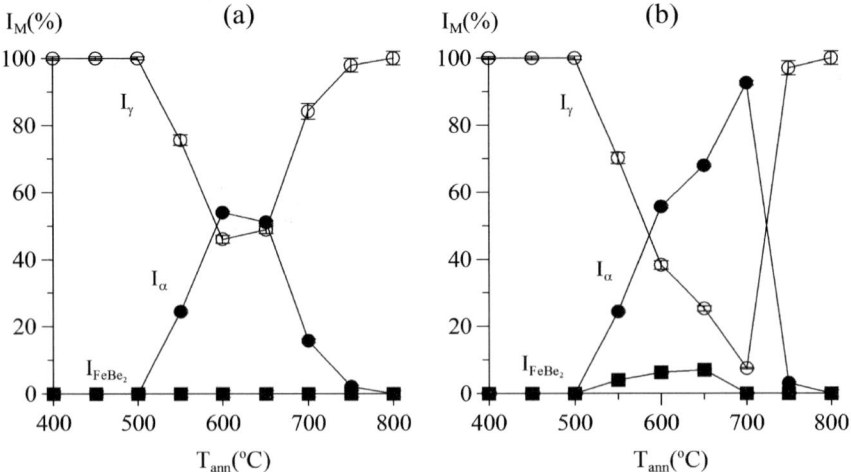

Figure 7 Relative intensities of partial MS-spectra for phases I_M in stainless steel samples covered with beryllium *versus* temperature of consequent isochronous annealings.

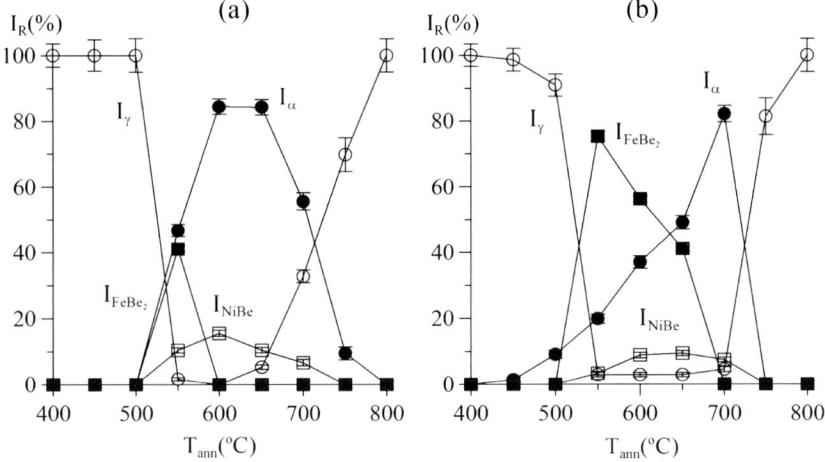

Figure 8 Relative intensities of phase reflexes for the angles $23.5° \leq 2\vartheta \leq 47.5°$ of the stainless steel samples at the beryllium coating side *versus* temperature of consequent isochronous annealings.

α-phase with its relative intensity achieved at the consequent annealing (600 °C). Further thermal treatment results in reverse $\alpha \rightarrow \gamma$-transformation. For the sample coated with $d_{Be} = 2$ μm there is also observed formation of α-phase at 550 °C with maximal value for relative intensity achieved at 700 °C. Moreover, at the temperature range 500 $\div$ 700 °C there has been found presence of the beryllide $FeBe_2$. Increase in annealing temperature results in recovery of the γ-phase.

Since we used Mössbauer spectroscopy in absorption geometry provides information on phase transformation in the sample bulk, there were performed X-ray phase investigations from both sides of the sample. As in case of mossbauer spectroscopy, both samples revealed $\gamma \rightarrow \alpha$ transformation and only a sample with $d_{Be} =$

 Springer

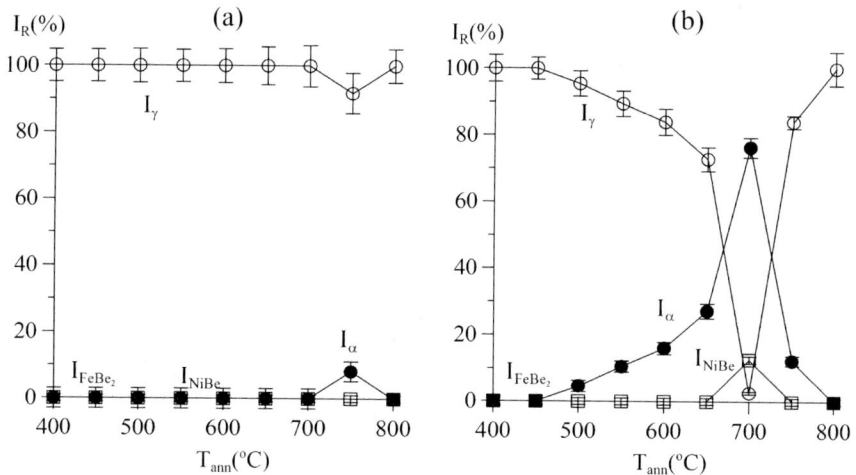

Figure 9 Relative intensities of phase reflexes for the angles $23.5° \leq 2\vartheta \leq 47.5°$ of the stainless steel samples from the backing side *versus* temperature of consequent isochronous annealings.

2 μm demonstrated presence of the beryllide $FeBe_2$. At the same time, observed $\gamma \rightarrow \alpha$-transformation is accompanied with formation of a phase without iron – Nickel beryllide NiBe (Figures 6, 7).

For more detailed analysis of X-ray diffractograms we have chosen the angle interval $23.5° \leq 2\vartheta \leq 47.5°$ where reflexes of all the revealed phases are presented at. Upon model decoding of selected fragments of X-ray diffractograms using SPECTR code [11, 12] were obtained annealing temperature dependencies for relative intensities I_R of phase reflexes from subsurface layers of the steel samples 12Cr18Ni10Ti coated with $d_{Be} = 1$ and 2 μm from the coating side (Figure 8) and the substrate (Figure 9). As one can see at Figure 8, subsurface layers of thin foils coated with beryllium are characterized by the same laws of phase transformations as are for its bulk. Formation of phases due to beryllium diffusion happens upon annealing at 550 °C. At that in the process of successive annealing relative intensity I_R for the newly formed α-phase increases reaching its maximum at $600 \div 650$ °C (for a sample coated with $d_{Be} = 1$ μm) and 700 °C (for a sample coated with $d_{Be} = 2$ μm). Further increase of annealing temperature results in martensite decomposition.

Let us pay attention to correlative changes in relative intensities of the nickel beryllide and α-phase. Let us also note that increase in coating thickness results in significant increase of relative intensity for iron beryllide reflexes and lower intensity of nickel beryllide reflexes.

Thermal annealing of $d_{Be} = 1$ μm steel foil did not result in any distinguishable phase transformations in/from the backing (see Figure 9), while for the sample $d_{Be} = 2$ μm there have been observed considerable changes: Intensive $\gamma \leftrightarrow \alpha$-transformations with simultaneous formation of nickel beryllide.

Comparative analysis of relative intensity dependencies for X-ray reflexes for various phases at subsurface layers of investigated beryllium steel foils showed important role that thickness of the beryllium coating plays in depth distribution of the phases.

Taking into account all stated above, one can make a conclusion that nickel beryllide is formed simultaneously with $\gamma\rightarrow\alpha$-transformation and, at sufficient beryllium concentration, there is formed iron beryllide provided that increase in beryllium concentration results in increase in $FeBe_2$ content. In its turn, nickel beryllide decomposition is accompanied with reverse $\alpha\rightarrow\gamma$-transformation. Thermally induced phase transformations in stainless steel coated with beryllium begin from the subsurface layer of a sample. At higher annealing temperatures and coating thicknesses this process involves deeper layers. At that the following sequence of phase transformations takes place:

1) first portions of beryllium atoms, being introduced into steel lattice, bind nickel atoms and form the chemical composition NiBe; its crystals split off the lattice and depletion of steel with nickel leads to $\gamma\rightarrow\alpha$-transformation;
2) further enrichment of α-solid solution with beryllium results in formation of iron beryllides.

4. Conclusions

The following can be concluded from the performed investigations of phase transformations in 12Cr18Ni10Ti stainless steel foils at thermal beryllization.

1. There have been determined sequences of phase transformations and relative amount of iron-containing phases at the surface and in the bulk of samples subjected to thermal beryllization.
2. It has been revealed that thermal beryllization of stainless steel thin foils results in $\gamma\rightarrow\alpha$ transformation and formation of the beryllides NiBe and $FeBe_2$.
3. It has been revealed that direct $\gamma\rightarrow\alpha$- and reverse $\alpha\rightarrow\gamma$-transformations are accompanied by, correspondingly, formation and decomposition of the berylliede NiBe.
4. It has been shown that distribution of formed phases in sample bulk is stipulated by local concentration of beryllium.
5. Based on obtained experimental data there has been proposed a physical model of phase transformations in stainless steel at thermal beryllization:

   – during initial diffusion of beryllium atoms there is formed the beryllide NiBe in the γ-phase what results in lowering of Ni atoms local concentration and transformation $\gamma\rightarrow\alpha$;
   – further diffusion and increase of Be-atoms concentration in α-phase at sufficient amount of beryllium atoms results in formation of the beryllide $FeBe_2$.

References

1. Kogan, B.I., Kapustinskaya, K.A., Topunova, G.A.: Beryllium. Nauka, Moscow (1975)
2. Knapton, A.G., West, K.B.C.: J. Nucl. Mater. **3**, 239 (1961)
3. Vickers, W.: In: The Metallurgy of Beryllium. Chapman & Hall, London (1963)
4. Hedge, E.S., et al.: In: Beryllium Technology, vol. 2. New York (1966)
5. Zemskov, G.V., Melynik, P.I.: Metallovedenie i termicheskaya obrabotka metallov **3**, 62 (1966) (in Russian)

6. Zbozhnaya, O.M., Borisov, E.V.: Fiziko-himicheskaya mekhanika materialov **10**, 64 (1974) (in Russian)
7. Altovsky, R.M., Vasina, E.A.: Atomnaya energia **38**, 333 (1975) (in Russian)
8. Altovsky, R.M., Panov, A.S.: Corrosion and Compatibility of Beryllium. Atomizdat, Moscow (1975) (in Russian)
9. Papirov, I.I.: Structure and Characteristics of Beryllium Alloys. Energoizdat, Moscow (1981) (in Russian)
10. Papirov, I.I.: Beryllium in Alloys. Energoatomizdat, Moscow (1986) (in Russian)
11. Rusakov, V.S.: Mossbauer Spectroscopy of Locally Non-Homogeneous Systems. Almaty (2000) (in Russian)
12. Nikolayev, V.I., Rusakov, V.S.: Mossbauer Studies of Ferrites. Moscow (1985) (in Russian)
13. Rusakov, V.S.: Izvestia AN. Seria fizicheskaya **63**, 1389 (1999) (in Russian)
14. Zemcik, T., Jakesova, M., Suwalski, J.: In: Proc. Intern. Conf. Appl. Mossbauer Effect, P.513. Jaipur, India (1982)
15. Ohta, K.: J. Appl. Phys. **39**, 2123 (1968)
16. Mints, R.I., Semyonkin, V.A.: Ukrainskii fizicheskii zhurnal **20**, 596 (1975) (in Russian)

Hyperfine Interact (2005) 164: 87–97
DOI 10.1007/s10751-006-9236-2

Mössbauer Spectroscopy of Locally Inhomogeneous Systems

V. S. Rusakov · K. K. Kadyrzhanov

Abstract The paper considers ways for obtaining information from Mössbauer spectra of locally inhomogeneous systems. The entire notion *locally inhomogeneous system* (LIS) is given a more precise definition applied to Mössbauer spectroscopy. There are considered factors that lead to local inhomogeneity of hyperfine interactions and its mechanisms. Application of LIS Mössbauer spectra processing and analysis methods are discussed. Ways for comprehensive utilization of various methods are described along with the role of *a priori* information at all processing stages.

Key words Mössbauer spectroscopy · local inhomogeneity · locally inhomogeneous system · spectra processing · a priory information.

1. Introduction

Substances with characteristic local inhomogeneities – with different from position to position neighborhood and properties of like atoms – gain recently increased scientific attention and wide practical application. We would call a system locally inhomogeneous if atoms in the system are in non-equivalent atomic positions and reveal different properties [1, 2]. Such systems are, first of all, variable composition phases, amorphous, multi-phase, admixture, defect and other systems. LIS are most convenient model objects for studies of structure, charge, and spin atomic states, interatomic interactions, relations between matter properties and its local characteristics as well as for studies of diffusion kinetics, phase formation, crystallization and atomic ordering; all that explains considerable scientific interest in such LIS.

V. S. Rusakov (✉)
M.V. Lomonosov Moscow State University, Moscow, Russia
e-mail: rusakov@moss.phys.msu.ru

K. K. Kadyrzhanov
Institute of Nuclear Physics NNC RK, Almaty, Kazakhstan

 Springer

Such systems find their practical application due to wide spectrum of useful, and sometimes unique, properties that can be controlled varying character and degree of local inhomogeneity.

Mössbauer spectroscopy is one of the most effective methods for investigation of LIS. Local character of obtained information combined with information on cooperative phenomena makes it possible to run investigations impossible for other methods. Mössbauer spectroscopy may provide with abundant information on peculiarities of macro- and microscopic state of matter including that for materials without regular structure. At the same time, analysis, processing and interpretation of Mössbauer spectra of LIS face considerable difficulties. Development of computer technique is accompanied with development of mathematical methods used for obtaining physical information from experimental data. The methods make it possible to improve considerably, with some available *a priori* information, effectiveness of the research. Utilization of up-to-date mathematical methods in Mössbauer spectroscopy requires not only adaptation of these methods for specific physical tasks and their software realization, but development of the methods for application and physical interpretation of obtained results.

Present paper considers ways for obtaining information from Mössbauer LIS spectra. Application of LIS Mössbauer spectra processing and analysis methods are discussed. Ways for comprehensive utilization of various methods are described along with the role of *a priori* information at all processing stages.

2. Locally inhomogeneous systems in Mössbauer spectroscopy

The entire notion *local inhomogeneity* directly relates to the state and, therefore, properties, of an atom at some position; these properties are first of all defined by the neighborhood and its characteristics. Characteristics of atomic neighborhood can be split up into the following categories (see Figure 1):

- *topological* characteristics (characteristics of spatial arrangement of neighboring atoms): Elements with point (local) symmetry, spacings and angles between the atoms, coordination numbers, radii of coordination spheres, catenation angles of polyhedrons, etc.;
- *composition (chemical)* characteristics of the neighborhood: Amount and arrangement of different type neighbor atoms;
- *physical* characteristics of the neighborhood (i.e. parameters that characterize state and interaction of neighboring atoms): Electronic configuration, spin, orbital and magnetic moments, etc.

External conditions such as temperature, pressure, field gradients and strengths, may influence considerably on both characteristics of atomic neighborhood and directly on physical characteristics of the atom at the given position (Figure 1). Characteristics of the atomic neighborhood entirely define equivalence or non-equivalence of the atomic positions. So, local non-uniformity in atomic properties can be classified based on its origin as *topological, composition, physical* and *combined* ones [2].

Since characteristics of the neighborhood and physical properties of an atom in a given position define hyperfine interactions (HI) of its nucleus (Figure 1), then using HI one can investigate local inhomogeneity of atomic properties and one can speak in

 Springer

Figure 1 The main factors determining local inhomogeneity of hyperfine interactions.

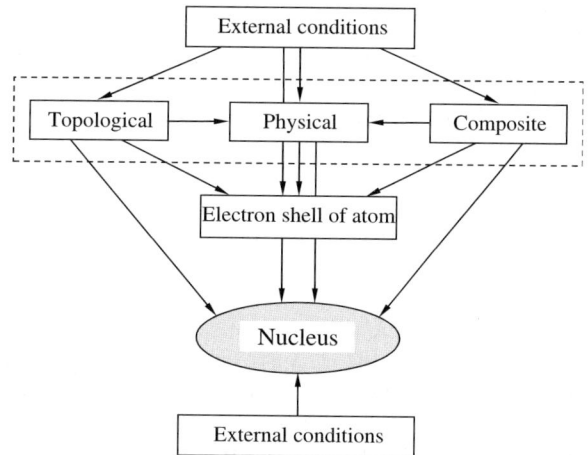

terms of *local inhomogeneity of HI* (LIHI). In this case classification of LIHI remains the same as for local inhomogeneity of atomic properties.

Each type of HI is associated with its hyperfine parameter of Mössbauer spectrum: Electric monopole – shift δ of Mössbauer line, electric quadrupole – constant of quadrupole interaction e^2qQ and asymmetry parameter η (or quadrupole shift ε of the spectrum components), magnetic dipole – effective magnetic field H_n in the region of nucleus.

Shift of the Mössbauer line δ is due to the isomer (chemical) δ_I and temperature δ_T shifts. The temperature shift δ_T is defined by vibration spectrum of the nuclei and the isomer shift δ_I – by the charge density of electrons in the nucleus' vicinity. At that electronic density on a nucleus is directly related to electronic configuration of an atom and to electronic structure of the system, including the character of the atoms' chemical bonds. Shift of the line δ is mainly defined by the direct neighbors of the atom. Influence from the atoms in higher-order coordination spheres is delivered, as a rule, through position and electronic state of the atoms in the first coordination sphere (see Table I).

Interrelation of the isomer shift δ_I and electron charge density in vicinity of a nucleus stipulates notable sensitivity of the δ line shift to topological and composition local inhomogeneities (Table I). In case of LIS there is change in spacings from the atoms in the first coordination sphere of replacement of some atoms in this sphere by other-type atoms from position to position. For instance, ^{57}Fe nuclei may reveal the following shift change with the change of distances in different systems: $\frac{\partial \delta}{\partial \ln r} \cong 1 \div 5$ mm/s, and with replacement of one atom – ~ 0.01 mm/s. Shift of the Mössbauer line δ makes it possible to identify valence state of atoms, obtain information regarding both phase composition and peculiarities of electronic and crystalline structures of LIS under investigation.

Quadrupole shift ε (splitting $\Delta = 2\varepsilon$) of the spectral hyperfine structure components takes place in non-uniform electric field that partially eliminates degeneration of the nuclear level on the magnetic quantum number. In general case the electric field gradient in vicinity of a nucleus is made by localized charges of ion bases of the neighboring atoms (lattice contribution), polarized conduction electrons and valence

🕮 Springer

Table I Sensitivity of hyperfine parameters

Hyperfine parameters		δ	ε	H_n
Number of coordination spheres		$1 \div 2$	$1 \div 10$	$1 \div 5$
Distance, Å		$2 \div 4$	$2 \div 20$	$2 \div 10$
The characteristics of an atom environment	Topological	+/−	+	+/−
	Composite	+/−	+	+
	Physical	−	+/−	+

"+" – high, "+/−" – appreciable and "−" – low sensitivity.

electrons of the Mössbauer atom. This means that the quadrupole shift ε is sensitive, first of all, to topological and composition local inhomogeneities (Table I).

Assessment of the change in ε due to lattice contributions only shows for ^{57}Fe that sensitivity region in case of quadrupole electric interaction is much wider than for the case of monopole electric interaction. This region has ~10 coordination spheres and can expand to the radius of up to 20 Å (Table I). In particular, appearance of an +3 ion at ~10 Å results in change of quadrupole shift for ~0.006 mm/s what is quite enough for revealing using modern methods for processing and analysis of Mössbauer spectra. If such ion happens to appear in the first coordination sphere, it results in huge ~0.8 mm/s change in quadrupole shift that exceeds its natural line width for eight times. Quadrupole shift ε describes peculiarities of electronic and crystalline structures of LIS and may provide with useful information both on symmetry of the closest vicinity of Mössbauer atoms and about their electronic configuration.

At investigations of magnetic-ordered systems, the most sensitive to local inhomogeneity among the Mössbauer spectra parameters is effective magnetic field H_n. This field is generated by both an atom and its neighbors. Effective field H_n is particularly sensitive to composition and physical (magnetic) local inhomogeneities (Table I) [1]. It is known that replacement of one magnetic atom with another one in the close vicinity may change the field at the ^{57}Fe nuclei for 20–30 kOe mainly due to change in Fermi contribution H_F. Assessment of dipole–dipole contribution H_{dip} into H_n shows that sensitivity region may cover ~5 coordination spheres and expand in this case up to 10 Å in radius (Table I). In particular, an atom with magnetic moment 5 μ_B produces the field of $H_{dip} \sim 0.09$ kOe at distance 10 Å what is quite enough for registration of the change in H_n, and the same atom produces the field of ~12 kOe at 2 Å. Data on effective magnetic field H_n may provide information on local peculiarities of atomic, magnetic and electrical structures of investigated magnetic-ordered LIS.

Presence of a considerable amount of to some extent non-equivalent positions of Mössbauer atoms (nuclei) in the investigated sample is typical for LIS. Spectra of such samples in form of a superposition from many subspectra may provide with various useful information on phase composition and peculiarities of crystalline, electronic and magnetic structures for each phase. There are, of course, some difficulties when extract this information. Complex structure of the spectra requires special spectrum analysis and processing methods involving modern mathematics and software. At that, choice of information extraction method becomes of importance. To considerable extent this choice is defined by the particular task based on available *a priori* information about the sample or other considerations.

 Springer

Table II Methods for analysis and processing of Mössbauer spectra realized within the software package MSTools

Analysis and processing method of Mössbauer spectra	A priori information			MSTools
	S	A	ST	
Resolution enhancement and noise suppression	+	−	−	RESOL, DISTRI
Model fitting	+	+	−	SPECTR
Distribution functions recovery	+	+	−	DISTRI
Comparative analysis with standard or modeling spectra	+	−	+	PHASAN
Simulation	−	−	−	RESOL, SPECTR, DISTRI, PHASAN

3. Classification of Mössbauer spectra analysis and processing methods

Application of Mössbauer spectroscopy for investigations of LIS requires solving of a set of tasks at analysis and processing of the Mössbauer spectrum. Content of the tasks depends both on a specific object under investigation and on *a priori* information the investigator has [2]. In its turn, each of these tasks implies that there are specific methods available for processing of the Mössbauer spectra and that the methods meet the specific requirements of this task.

Let us consider specific tasks related to utilization of Mössbauer methods in investigations of LIS and correspondingly to them classify the Mössbauer spectra analysis and processing methods. We provide this classification with examples of the methods realized by us within the software package MSTools [2] designed for effective applications in Mössbauer spectroscopy LIS (see Table II).

In some cases of Mössbauer investigations of LIS it is necessary to improve spectrum quality by higher resolution or effective noise suppression. As a rule, such necessity arises with lack of *a priori* information about the object. Statistical noise in experimental spectrum or limited line width of the radiation source o_s may somehow 'hide' peculiarities of the spectrum with desired information. The result of solving such task is a new, transformed spectrum with considerably increased resolution or effectively suppressed noise (Table II).

Solving such task within the software package MSTools we realized linear methods for processing of Mössbauer spectra – filtration, regularization and 'discrepancy' [2]. A characteristic feature of these methods is absence of any valuable *a priori* information AI(A) about the object under investigation. But effectiveness and reliability of the processing depend greatly on reliability and completeness of *a priori* information AI(S) about the radiation source and peculiarities of the spectrometer hardware performance. This processing method has been realized in a program RESOL [2]. Spectrum quality may be improved with recovery of distribution function $p(v)$ for the single resonance line shift (program DISTRI). In this case recovery of the function $p(v)$ can be interpreted as a transformed spectrum of better quality (for more details see [2]).

Another task related to investigation of LIS is the model fitting of spectra (Table II). This implies search for or specification of values for comparatively narrow set of physical parameters that, within the chosen model, uniquely describe the

 Springer

Mössbauer spectrum. When applied to LIS such task statement is reasonable when two conditions below are met:

1. a researcher possesses (or thinks that possesses) sufficient *a priori* information about both the spectrometer AI(S) and the object AI(A), first of all, about its phase composition, atomic distribution, point symmetry, chemical bounds and valence;
2. number of independent parameters that describe spectrum within the chosen model is not too high (i.e. much less than the number of points in the experimental spectrum).

Such situation may take place if the substance under investigation is of regular crystalline and magnetic structure and the sample is homogeneous in its composition. The processing method is based on the least squares method to get the best fit for description of the experimental curve and superposition of so-called non-rigid bounds; it is also based on superposition of subspectra valid for 'thin' enough samples. The method has been realized in the program SPECTR [2] and designed for obtaining of qualitative information on hyperfine interaction parameters for Mössbauer nuclei.

Experimenters quite frequently deal with LIS of amorphous substances, variable composition phases of doped systems. Description of the system state with some discrete set of physical parameters' values is difficult then as well as of Mössbauer nuclei in this system. When process Mössbauer spectra of such systems it is necessary to recover distribution functions of the spectrum parameters. This problem has been solved in the program DISTRI using regularization method in its iterations variant [1, 2]. This variant of the method makes it possible to put physically grounded conditions both onto the values of recovered distribution functions and other variable using *a priori* information about the spectrometer AI(S) and the object under investigation AI(A). As a result there are obtained distribution functions for subspectra parameters with evaluation of statistical errors and characteristics of these functions.

The next considered task implies comparison of Mössbauer spectrum obtained for a sample with spectra of standard samples. This task is of importance at analysis of multiphase systems when investigator has no information on each separate phase and it is necessary to make qualitative or quantitative phase analysis only. Almost no *a priori* information on a spectrometer AI(S) and a sample AI(A) is required in this case, but one needs spectra of standard samples with low noise levels, i.e. high-quality spectra. The core of the method is in searching for optimal combination of the standard samples spectra for description of the spectrum for the sample of interest; the method is based on application of the least squares method and superposition principle. Upon processing of the spectrum it is possible to obtain both weight (absolute and relative) and isotopic (absolute) content of the standards in the sample of interest. Within MSTools software the method is realized in the program PHASAN [2] and is designed for qualitative and quantitative spectra analyses independently on availability of information on their structure. This method of spectra analysis can also be used for revealing of slight changes in a spectrum at any directed impact on the sample such as heating, implantation, laser annealing, ageing, composition change and so on.

The last but still very important task is simulation of Mössbauer spectra with the following comparison it with experimental spectrum. It is possible to simulate Mössbauer spectra with any of the programs SPECTR, DISTRI or PHASAN, but only within the framework of the models realized by these programs.

 Springer

As we can see, the considered methods realized within MSTools software package cover almost all possible types of analysis and processing of Mössbauer spectra excepting only the case with relaxation effects. Each of these methods is most effective when dealing with a specific task it has been designed for, but they can also be used for other tasks adding and 'helping' each other.

4. Comprehensive approach to processing and analysis of spectra

As a rule, at Mössbauer study of a specified LIS sample not one, but several from considered above tasks for processing of spectra emerge.

First, a specific case rarely makes it possible to assign the object to a specific substance class. For instance, LIS under investigation may have regular crystalline and non-regular magnetic structures at the same time. Or, in some cases, variable composition phases may reveal their properties (with corresponding change in the Mossbauer spectra) as a small, but still existing complex of regular structures. In all these cases it is reasonable to set both tasks of model fitting and recovery of the distribution function for spectrum parameters.

Second, peculiarities of the considered spectra processing methods make it possible to use these methods solving other tasks they were not designated to. For example, in order to obtain weight and isotopic content of some phases in LIS sample one can use model fitting of the spectrum. One can also try to describe a Mössbauer spectrum, for instance, of a multi-phase sample with regular phase structures by recovering distribution functions for subspectra parameters. Or *vice versa*, spectrum of a sample with non-uniform structure can be described with a large amount of subspectra using model fitting.

Third, purposeful consequent utilization of various methods may be an advantage. Such approach is particularly effective when almost no *a priori* information is available about a LIS sample and its Mössbauer spectrum revealed complex multi-component structure. The need in simultaneous solution of several tasks related to analysis and processing of Mössbauer spectra calls for comprehensive utilization of the methods. Processing of a spectrum with several methods independently from one side adds up obtained information and, from the other side, improves reliability of the conclusions made.

As one can see, different methods for analysis and processing of Mössbauer spectra leave the possibility for their close interaction while application of a specific method, its effectiveness depends considerably on the results of application of other methods. Such interaction can be realized using information provided by one of the methods as *a priori* information for another one. Moreover, utilization of any of the methods is impossible without utilization *a priori* information in some form and to certain extent.

5. Role of *a priori* information

As early as at the stage of task formulation and choice of spectrum processing method, availability of *a priori* information, its character and completeness play crucial role. Really, information on crystalline and magnetic structures of a sample, short

or long range ordering there, on isomorphic replacement of atoms, on admixtures in the sample may set up a task either of model fitting of the spectra, or of recovery of the parameter distribution functions. Even not high-quality and very reliable information on sample composition at real possibility to obtain standard spectra frequently explains the choice to compare the spectrum with spectra of such standard samples. Unavailability of *a priori* information about the object under investigation or its penury (even when this is accompanied by poor quality of experimental spectrum) forces to improve first the spectrum quality in order to obtain additional *a priori* information.

The same important role *a priori* information plays at direct application of the method for processing within a specific task. Reliable *a priori* information eliminates in some cases ambiguity of the processing result and improves its reliability.

In order to improve spectrum quality the information on real source and spectrometer AI (S) used is of importance: Form and width of radiation line, spectrometer performance quality (drifts in its electronic units, precision of continuous form of change in Doppler velocity v), about geometry of the experiment (angular aperture of the registered beam of γ-quanta and relative amplitude of the source oscillations).

In model fitting of spectra, utilization of *a priori* information on the investigation object AI(A) makes it possible to reduce dramatically the number of physically reasonable models within which the physical values of the parameters that describe the spectrum; these parameters help not only to identify the revealed phases or non-equivalent positions of atoms, but obtain new physical information. In such case very diverse information both qualitative (for instance, about magnetic structure colinearity, about the mechanism of exchange interactions, or presence of texture) and quantitative (number of non-equivalent positions, characteristic values of spectrum hyperfine parameters, specific cation distribution, orientation of easy direction, electronic configuration, etc.) can be used.

At recovery of distribution functions $p(z)$ of one of the spectrum parameters z, besides information on the source and the spectrometer AI(S), there is of importance any information about sample under investigation that enables making grounded guesses regarding the 'behavior' of other spectrum parameters and, in particular, about their possible correlation with z. This can be information on the mechanism of hyperfine interactions, chemical bonds or, for instance, presence of spontaneous striction, about interrelation of cation distribution with the symmetry of closest vicinity of the Mössbauer nucleon, and so on.

Information on standard samples AI(ST), about their possible presence in the investigated sample, about the spectrometers where spectra were obtained are the *a priori* information required for effective utilization of the method that compares spectrum of a sample under investigation with spectra of standard samples.

Peculiarities of the methods for analysis and processing of Mössbauer spectra assure in case of LIS utilization of not only each method independently, but also their combination. Role of *a priori* information becomes crucial at that. This interrelation is based on a quite obvious fact that new information obtained by any of the methods may be to certain extent used both for formulation of another specific task for analysis and processing, and for solving the same task with another method.

Therefore, comprehensive approach to processing of Mössbauer spectra is an interrelated consequent application of various methods with specific combination of these methods defined, first of all, by available a priory information before and after application of each of the individual methods.

 Springer

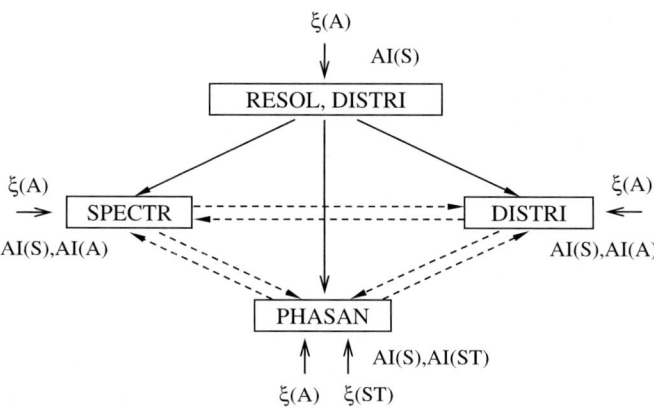

Figure 2 Possible functional interrelation of various methods.

6. Interrelation of various analysis and processing methods

Let us consider in more details the interrelation of various methods for analysis and processing of Mössbauer spectra at investigations of LIS (see Figure 2). A method of spectrum quality improvement (RESOL, DISTRI) takes a particular place among the methods for analysis and processing. In this method, different to all other, no *a priori* information about an object under investigation AI(A) is used. Moreover, this method has been designed for obtaining of such *a priori* information by increasing resolution in a spectrum or by effective noise suppression. The entire quality improvement may result in completion of the LIS investigation. However, in reality the results of the method are very frequently used in application of other methods at consequent processing stages. Analysis of the transformed spectrum (or distribution function $p(v)$) makes it possible to define with some reliability the number of lines in the spectrum n_l and reveal the number of subspectra n_s (i.e. number of supposed phases in the sample or of entirely non-equivalent positions of Mössbauer atoms). Results of such analyses can be used for creation of a specific model for consequent model fitting or recovery of distribution functions for the spectrum parameters.

As a rule, revealed values n_l and n_s together with *a priori* information AI(A) obtained from other experimental methods or due to gained experimental experience, makes it possible to choose between the other three processing methods. If it is known that a sample is of regular structure and supposed number of components is not high, it is necessary to formulate a task for model spectrum fitting (SPECTR) involving all the available *a priori* information into the constructed model. If spectrum quality improvement indicated non-uniform character of crystalline or magnetic structures (when it was not known *a priori*), one should set a task for recovery of spectrum parameter distribution functions (DISTRI) using available *a priori* information to set corresponding conditions. In case when revealed values of spectrum parameters indicate possible presence in the sample of at least one standard sample, it is worth to complete the task of comparison with standard sample spectra $\xi(ST)$ (PHASAN).

 Springer

In the most simple cases solving of one of these three last spectrum processing tasks results, as a rule, in completion of the analysis and processing. If investigator has got complete enough *a priori* information about the sample or sample spectrum is of high quality, then need in spectrum quality improvement is eliminated and one can simply use one of the programs SPECTR, DISTRI or PHASAN.

However, there might be a situation when none of the considered methods used individually provides reliable analysis of LIS spectrum. This may be stipulated by wrong method choice at previous stage (due to incompleteness of failure of *a priori* information AI(A)) or by the complexity of the sample under investigation that may contain, for instance, phases with considerably different structure regularity. If, for example, reconstructed distribution function $p(z)$ can be characterized by a set of local maxima, then, using this result as *a priori* information AI(A), one could set a task of model fitting. While if the function $p(z)$ is of a form characteristic for some of the available standard samples, then comparison with the spectra of standard samples would be an advantage.

Similar situation may happen at initial application of the model fitting of spectra. For example, close found optimal spectrum parameters and the natural desire to increase the number of components n_1 in the spectrum for considerable decrease of the functional χ-square or unreasonably large widths of the spectrum components led to formulation of a task to find distribution functions for corresponding parameters. From the other side, if the values of at least a part of the found parameters correspond well to a spectrum of at least one available standard sample, then the chance to solve the task of comparison with the spectra of standard samples should be realized.

In its turn, comparison with spectra of standard samples may lead to the conclusion that not all the spectrum of investigated sample is described by this set of spectra of standard samples. In this case depending on available *a priori* information it becomes very natural to run model fitting of the non-described part of the spectrum or to recover distribution functions for the parameters of this part.

7. Conclusions

Effectiveness of Mössbauer spectroscopy for investigation of local inhomogeneity systems depends greatly on adequacy of our physical understanding and adequacy of information extraction ways to the object of investigation. All the experience gained by the authors for the last years [3–17] in investigation of LIS showed fruitfulness of the presented above conception of hyperfine interaction local inhomogeneity, reasons for its appearance and mechanisms of its formation as well as high effectiveness of the comprehensive approach to processing and analysis of Mössbauer LIS spectra using *a priori* information.

References

1. Rusakov, V.S.: Izv. Ross. Akad. Nauk Ser. Fiz. **63**(7), 1389 (1999)
2. Rusakov, V.S.: Mössbauer spectroscopy of locally inhomogeneous systems. - INP NNC RK. Almaty, p. 431 (2000)
3. Rusakov, V.S., Kotelnikova, A.A., Bychkov, A.M.: Geokhimiya **11**, 1234 (1999)
4. Rusakov, V.S., Kadyrzhanov, K.K., Turkebaev, T.E.: Poverkhnost' **4**, 27 (2000)
5. Rusakov, V.S., Kupin, Yu.G., et al.: Geochem. Int. **38**(Suppl.3), S383 (2000)

6. Kadyrzhanov, K.K., Rusakov, V.S., Turkebaev, T.E.: Nucl. Instrum. Methods Phys. Res., B **170**(1–2), 85 (2000)
7. Kadyrzhanov, K.K., Rusakov, V.S., et al.: Nucl. Instrum. Meth. Phys. Res., B **174**, 463 (2001)
8. Kadyrzhanov K.K., Rusakov, V.S., et al.: Izv. Ross. Akad. Nauk. Ser. Fiz. **65**(7), 1022 (2001)
9. Kadyrzhanov, K.K., Rusakov, V.S., et al.: Hyperfine Interact. **141–142**(1–4), 453 (2002)
10. Rusakov, V.S., Chistyakova, N.I., Kozerenko, S.V.: Hyperfine Interact. (Ñ) **5**, 461 (2002)
11. Kadyrzhanov, K.K., Kerimov, E.A., et al.: Poverkhnost' **8**, 74 (2003)
12. Chistyakova, N.I., Rusakov, V.S., et al.: Material research in atomic scale by mossbauer spectroscopy. NATO Sci. Ser. II. **94**, 261 (2003)
13. Kadyrzhanov, K.K., Rusakov, V.S., et al.: Poverkhnost' **7**, 75 (2004)
14. Chystyakova, N.I., Rusakov, V.S., et al.: Hyperfine Interact. **156–157**(1–4), 411
15. Kadyrzhanov, K.K., Rusakov, V.S., et al.: Hyperfine Interact. **156–157**(1–4), 623
16. Rusakov, V.S., Kadyrzhanov, K.K., et al.: Poverkhnost' **12**, 22 (2004)
17. Rusakov, V.S., Kadyrzhanov, K.K., et al.: Poverkhnost' **1**, 60 (2005)

Hyperfine Interact (2005) 164: 99–104
DOI 10.1007/s10751-006-9239-z

Mössbauer Study of Ferrite-garnets as Matrixes for Disposal of Highly Radioactive Waste Products

**V. S. Rusakov · V. S. Urusov · R. V. Kovalchuk ·
Yu. K. Kabalov · S. V. Yudincev**

Abstract Mössbauer study of synthesized ferrite-garnet samples containing Zr, Th, Ce and Gd of the following composition: $1C - Ca_{2,5} Ce_{0,5} Zr_2 Fe_3 O_{12}$, $2C - Ca_{1,5} GdCe_{0,5} ZrFeFe_3 O_{12}$, $1T - Ca_{2,5} Th_{0,5} Zr_2 Fe_3 O_{12}$ and $2T - Ca_{1,5} GdTh_{0,5} ZrFeFe_3 O_{12}$ are carried out. As a result of ^{57}Fe Mössbauer study it is found that iron atoms in all investigated samples of garnets are in a trivalent state. The analysis of experimental Mössbauer spectra definitely specifies a various structural state of iron atoms in two investigated groups of samples: 1T, 1C and 2T, 2C. X-ray study have shown that 1T and 1C garnet samples crystallize in tetragonal space group $I4_1/acd$, but 2T and 2C samples crystallize in cubic space group $Ia3d$.

Key words Mössbauer spectroscopy · ferrite-garnet · actinide · subspectra · distribution function.

1. Introduction

Among the most promising matrixes for disposal of actinide-containing highly radioactive waste products are phases of the composition $A_3B_2T_3O_{12}$ with garnet structure. Bi- (Ca, Mn, Mg, Fe, Co, Cd) and trivalent (Y, REE) cations can be in dodecahedral A-site. Octahedral B-site can be occupied by bi-, three- (Fe, Al, Ga, Mn, In, Sc, Co, V), quadri- (Zr, Ti, Sn, Ru) and even pentavalent (Nb, Ta, Sb) cations. In the center of tetrahedron TO_4 there are usually quadrivalent (Si, Ge, Sn) cations, but also there are can be tri- (Al, Ga, Fe) and pentavalent (N, P, V, As) cations.

The content of actinide elements in natural garnets is insignificant (a share of percent). Therefore, natural garnet study does not give us the possibility to evaluate

V. S. Rusakov (✉) · V. S. Urusov · R. V. Kovalchuk · Yu. K. Kabalov
M.V. Lomonosov Moscow State University, Moscow, Russia
e-mail: rusakov@moss.phys.msu.ru

V. S. Urusov · S. V. Yudincev
Institute of Geology and Mineralogy of Ore Deposits, RAS, Moscow, Russia

dissolution of actinide elements in structure of these garnets. This information can be received only under investigations of the synthesized samples. The occupations of B- and T-sites by relatively large and low-valence cations give the possibility for actinide element entering into A-site. Fe^{3+} ion is the largest trivalent cation that can occupy the octahedral B- and tetrahedral T-sites simultaneously. Therefore, namely ferrites have largest sizes of frame consisting of octahedrons FeO_6 and tetrahedrons FeO_4 and so there are maximum dimensions of dodecahedrons AO_8.

2. Method of experiment

For studying the structure and properties of ferrite-garnet samples containing Zr, Th, Ce and Gd Mössbauer study of synthesized samples of the following composition: $1C - Ca_{2,5} Ce_{0,5} Zr_2 Fe_3 O_{12}$, $2C - Ca_{1,5} GdCe_{0,5} ZrFeFe_3 O_{12}$, $1T - Ca_{2,5} Th_{0,5} Zr_2 Fe_3 O_{12}$ and $2T - Ca_{1,5} GdTh_{0,5} ZrFeFe_3 O_{12}$ were performed. Since ionic radii values of Th^{4+} and Ce^{4+} are comparable to those of Pu^{4+}, Np^{4+}, U^{4+}, Am^{3+} and Cm^{3+} it will be reasonable to assume that ferrite-garnets with content of these actinide elements also can be obtained experimentally.

Investigations were carried out at room temperature using constant acceleration Mössbauer spectrometer MS1101E and source of ^{57}Co in Rh matrix. Analysis and processing of the spectra was performed using model fitting [1, 3] (program SPECTR) and calculation of hyperfine parameters distribution functions for subspectra [2, 3] (program DISTRI). These programs are included in MSTools software [3].

3. Experimental results and discussion

Mössbauer spectra of the investigated samples are the paramagnetic type spectra witch consist of several quadrupole doublets. Subspectra have essentially different values of quadrupole shift ε of spectrum components and values of Mössbauer line shift δ. Distribution functions $p(v)$ of single resonance line position on a scale of Doppler velocities v for all experimental spectra have been calculated for the finding of resonance line numbers in the spectrum and their belonging to subspectra. In addition it was accepted that natural line width is $\Gamma_\tau \cong 0.097$ mm/s, source line width $- \Gamma_s \cong 0.11$ mm/s according to it passport. Analysis extracted distribution functions $p(v)$ has allowed us to choose models and initial values of variable parameters. It were used for the further calculation of quadrupole shift distribution functions $p(\varepsilon)$ for each of subspectrum and model fitting of all experimental spectra. The model fitting was carried out in the assumption of presence of two independent asymmetric quadrupole doublets with components of equal intensity for most intensive subspectra. This model was decided on because the investigated garnets are phases of variable composition and are referred to so-called locally inhomogeneous systems [3]. For the same reason searching of linear correlation between quadrupole shift ε and Mössbauer line shift δ was carried out for these spectra during the calculation of quadrupole shift distribution functions $p(\varepsilon)$ [2].

The results of all stages of processing and the analysis of spectra for 1C, 2C and 1T, 2T garnets samples are shown in Figures 1, 2. Single resonance line position distribution function $p(v)$ for 1C sample definitely indicate the presence of three subspectra. Two quadrupole doublets have most intensity with approximately equal

Springer

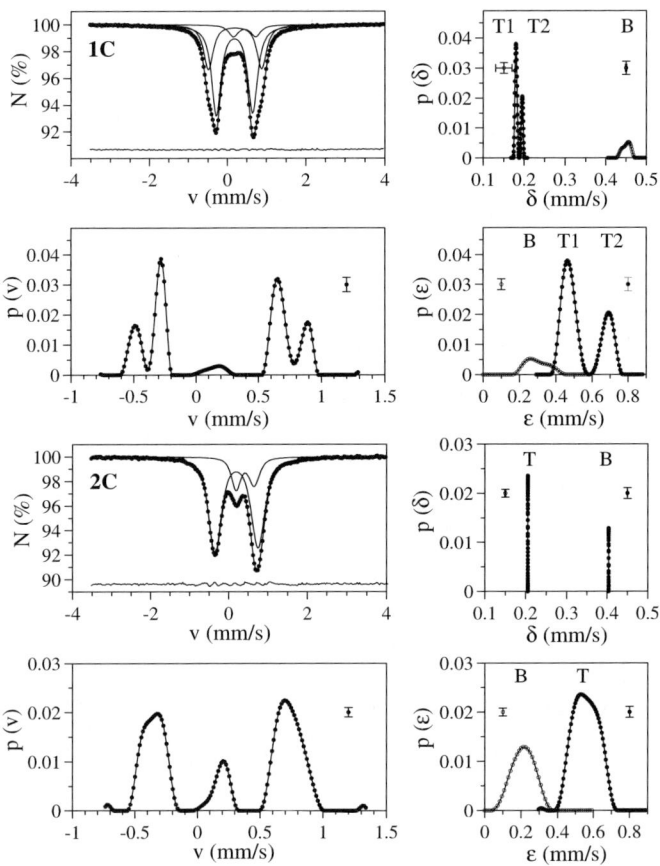

Figure 1 Results of model fitting and distribution functions calculation of a single resonant line position $p(v)$ and quadrupole shift $p(\varepsilon)$ for Mössbauer spectra of ferrite-garnets: 1C – $Ca_{2,5}Ce_{0,5}Zr_2Fe_3O_{12}$ and 2C – $Ca_{1,5}GdCe_{0,5}ZrFeFe_3O_{12}$.

Mössbauer line shifts δ and the third have essentially less intensity with great value of shift δ. Distribution functions $p(v)$ for 2C sample has shown the presence of only two well-defined subspectra with essentially various values of intensity I, shift δ and quadrupole shift ε (Figure 1). Similar results have been obtained for other two samples of the ferrite-garnets containing Th – for $Ca_{2,5}Th_{0,5}Zr_2Fe_3O_{12}$ (1T) and $Ca_{1,5}GdTh_{0,5}ZrFeFe_3O_{12}$ (2T) (Figure 2).

Values of relative intensity I and all hyperfine parameters for experimental spectra were obtained using calculation of distribution functions $p(\varepsilon)$ and model fitting of spectra (see Table I). The intensities ratio of subspectra is equal to the ratio of positions populations corresponding to these subspectra under the assumption that values of Debye temperatures describing vibration spectra of Mössbauer atoms in these positions are close to each other.

Subspectra with smaller values of shift $\delta \sim 0.18 \div 0.25$ mm/s and large values of quadrupole shift $\varepsilon \sim 0.47 \div 0.69$ mm/s for each experimental spectrum (Table I) can be attributed to the trivalent iron ions. These ions exist in a high-spin state in

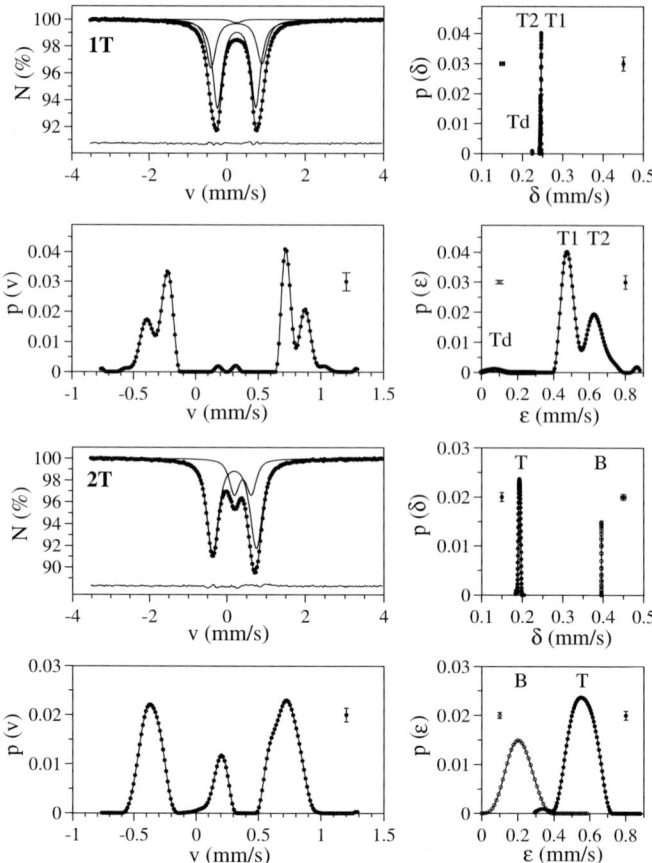

Figure 2 Results of model fitting and distribution functions calculation of a single resonant line position $p(v)$ and quadrupole shift $p(\varepsilon)$ for Mössbauer spectra of ferrite-garnets: 1T – $Ca_{2.5}Th_{0.5}Zr_2Fe_3O_{12}$ and 2T – $Ca_{1.5}GdTh_{0.5}ZrFeFe_3O_{12}$.

tetrahedral oxygen environment in crystallographic T-site in structure of the garnet (compare with results of papers [4, 5]). The quadrupole doublets with essentially larger values of shift $\delta \sim 0.40 \div 0.45$ mm/s and smaller quadrupole shift $\varepsilon \sim 0.21 \div 0.29$ mm/s correspond to Fe^{3+} ions in a high-spin state in octahedral oxygen coordination in crystallographic B-site.

As a result of ^{57}Fe Mössbauer study it is found that Fe atoms in all investigated samples of garnets are in a trivalent state. Two various tetrahedral sites (T1 and T2) of iron ions are defined reliably in 1T and 1C samples. The intensities ratio of subspectra corresponding to these sites is nearly equal to 2: For 1T – $1.6 \div 1.8$ and for 1C – $2.0 \div 2.1$ (see Table I). In addition the iron ions in octahedral B-site are found in 1C sample ($\sim$8%) and not in 1T sample. The quadrupole doublet with small intensity $I \cong$ 1% which is present at a spectrum of 1T sample can be to attribute to Fe^{3+} ions. These ions are probably located in tetrahedral sites (Td) of structure of the not determined by us phase.

 Springer

Table I Results of model fitting (SPECTR) and quadrupole shift distribution functions (DISTRI) analysis for Mössbauer spectra of investigated garnet samples (the standard deviation of statistical mistakes of the corresponding parameter is indicated in brackets)

Sample	Site	SPECTR				
		I, %	*δ, mm/s	ε, mm/s	Γ_1, mm/s	Γ_2, mm/s
1T	T1	63.8(6)	0.248(2)	0.486(2)	0.288(3)	0.291(3)
	T2	35.1(6)	0.243(2)	0.648(2)	0.288(3)	0.313(4)
	Td	1.1(1)	0.238(12)	0.062(12)	0.220(20)	
1C	T1	61.2(6)	0.181(2)	0.468(2)	0.291(2)	0.301(3)
	T2	30.5(5)	0.193(2)	0.686(2)	0.291(2)	0.296(4)
	B	8.3(3)	0.439(11)	0.286(11)	0.304(13)	
2T	T	76.2(5)	0.192(1)	0.556(1)	0.349(2)	0.374(3)
	B	23.8(5)	0.404(3)	0.215(3)	0.303(5)	
2C	T	80.1(9)	0.201(2)	0.552(2)	0.382(3)	0.409(6)
	B	19.9(9)	0.422(6)	0.231(6)	0.285(6)	0.326(12)

Sample	Site	DISTRI			
		I, %	*δ, mm/s	ε, mm/s	$\Gamma_{p(\varepsilon)}$, mm/s
1T	T1	60.0(1.5)	0.246(2)	0.474(15)	0.084(5)
	T2	38.8(1.9)	0.245(3)	0.626(20)	0.109(21)
	Td	1.6(3)	0.224(10)	0.071(12)	0.097(5)
1C	T1	62.6(1.5)	0.180(4)	0.463(18)	0.096(6)
	T2	30.3(1.2)	0.195(6)	0.689(21)	0.086(8)
	B	7.1(6)	0.451(60)	0.262(32)	0.176(36)
2T	T	78.1(9)	0.193(9)	0.551(26)	0.195(8)
	B	21.9(9)	0.396(3)	0.205(18)	0.169(10)
2C	T	81.5(9)	0.206(3)	0.531(29)	0.212(6)
	B	18.5(9)	0.404(4)	0.217(27)	0.170(12)

*Here and farther Mössbauer line shift δ represent concerning to α-Fe.

Unlike 1T and 1C samples the iron ions in 2T and 2C samples occupy only one crystallographic tetrahedral T-site in essentially non-uniform local environment. And the significant amount of Fe atoms exists in octahedral B-site. Relative intensity of subspectrum corresponding to B-site for 2T sample were found 22 ÷ 24% and for 2C sample – 19 ÷ 20%. In addition the line widths and widths $\Gamma_{p(\varepsilon)}$ of quadrupole shift distribution functions $p(\varepsilon)$ for T-site are essentially exceed those for B-site (see Table I). This circumstance is caused by strong non-uniformity of local environment of iron ions in tetrahedral T-site.

Thus, the Mössbauer data indicate various structural states of iron atoms in two investigated groups of samples: 1T, 1C and 2T, 2C. Two tetrahedral T1- and T2-sites of Fe^{3+} populated in the attitude ~2:1 are found in the first pair of samples. There is only one tetrahedral T-site of iron atoms in the second pair of samples and a significant amount of Fe atoms occupies octahedral B-site.

It is necessary to note that the splitting of tetrahedral sites in ratio ~2:1 in ferrite-garnets with similar structure was observed by authors of recently published paper

 Springer

[6]. They have explained unusual behaviour of garnets of such structure by a non-uniform local environment of two different kinds of tetrahedral site of iron atoms. However, it is obviously that local cation environment of T-site should have random character under the random distribution of atoms substituting each other in A- and B-sites of structures. It can lead only to lines broadening of one quadrupole doublet, but not to the spectrum splitting in two doublets. Such splitting testifies to high degree of short range ordering that it should be consequence of an establishment of the long range ordering in structure. Hence, 1T and 1C samples can have different crystal structure in comparison with samples 2T and 2C.

From these considerations X-ray study with application of a full profile analysis method have been carried out. It is shown, that 1T and 1C garnets samples crystallize in tetragonal space group $I4_1/acd$ with parameters of unit cells: $a = 12.7546(3)$, $c = 12.7588(3)$ Å and $a = 12.7130(3)$ Å, $c = 12.7153(5)$ Å correspondingly. 2T and 2C samples crystallize in cubic space group $Ia3d$ with parameters of unit cells: $a = 12.622(5)$ and $a = 12.660(5)$ Å, accordingly. Also it was found that in 1T and 1C samples there was a small amount of perofskite fraction of $CaZrO_3$ composition: 3% (1T) and 8% (1C).

As a result of performed investigations there were revealed idealized crystal-chemical formulas of synthesizes samples with tetragonal space group $I4_1/acd$:

$$1T - (Ca_{1.50}Th_{0.50})^{A1}(Ca_{0.90}Th_{0.10})^{A2}Zr_{2.00}(Fe_{1.93})^{T1}(Fe_{1.00})^{T2}O_{12},$$
$$1C - (Ca_{1.50}Ce_{0.50})^{A1}(Ca_{0.90}Ce_{0.10})^{A2}Zr_{2.00}(Fe_{1.93})^{T1}(Fe_{1.00})^{T2}O_{12}.$$

References

1. Nikolaev, V.I., Rusakov, V.S.: Mössbauer study on ferrites. Izd-vo Mosk. University, Moscow (1985). 224p. [in Russian]
2. Rusakov, V.S.: Izv. Ross. Akad. Nauk. Ser. Fiz. **63**, 1389 (1999), [in Russian]
3. Rusakov, V.S.: Mössbauer Spectroscopy of Local Inhomogeneous Systems. Almaty, INP NNC of Kazakstan (2000). 431p. [in Russian]
4. Menil, F.: J. Phys. Chem. Solids **46**, 763 (1985)
5. Parish, R.V.: Mössbauer spectroscopy and the chemical bond. In: Dickson, D.P.E., Berry, F.J. (eds.) Mössbauer Spectroscopy. Cambridge University Press, Cambridge (1986). London, New York, New Rochelle, Melbourne, Sydney. 274p.
6. Shabalin, B.G., Polshin, E.V., Titov, Yu.O., Bogacheva, D.O.: Miner. Zh. **25**, 41 (2003)

Hyperfine Interact (2005) 164: 105–109
DOI 10.1007/s10751-006-9238-0

Magnetic Phase Transitions in Nanoclusters and Nanostructures

I. P. Suzdalev · Yu. V. Maksimov

Abstract New phenomena – the first order magnetic phase transitions were observed in nanoclusters and nanostructures. For isolated ferrihydrite nanoclusters ($d \sim$ 1–2 nm) in porous materials, for α-,γ-Fe$_2$O$_3$ nanoclusters ($d \sim$ 20–50 nm) and for composites of nanostructured metallic Eu with additives of α-, γ-Fe$_2$O$_3$ nanoclusters and adamantane the critical temperatures (T_C, T_N) and magnetic cluster critical sizes (R_{cr}) were determined by means of thermodynamic models and Mössbauer spectroscopy. The first order magnetic phase transitions (jump-like) proceed by such a way when magnetization and magnetic order disappear by jump without superparamagnetic relaxation. According to thermodynamic model predictions the cluster and interface defects were suggested to play the main role in magnetic behavior. Thus, for the defective α-, γ-Fe$_2$O$_3$ nanoclusters, at $R \leq R_{cr}$, the presence of the first order (jump-like) magnetic phase transition was described in terms of magnetic critical size of cluster. The action of high pressure (up to 2 GPa) with shear (120–240°) was effective for defect generation and nanostructure formation. For nanosystems including iron oxide nanoclusters, adamantane and metallic europium and subjected to shear stress under high pressure loading the critical value of defect density was estimated by the study of the character of magnetic phase transition. First-to-second-order (nanostructured metallic Eu) and second-to-first-order (α-, γ-ferric oxide nanoclusters) changes of the character of magnetic phase transition were shown to accompany by the variation of critical temperatures compared to the corresponding bulk values.

Key words α-, γ-ferric oxide nanoclusters · nanostructured metallic Eu · magnetic phase transitions · defects · thermodynamics · Mössbauer spectroscopy.

The first order magnetic phase transitions were observed in nanoclusters and nanostructures [1, 2]. These phenomena are characterized by the jump like disap-

I. P. Suzdalev (✉) · Yu. V. Maksimov
N.N. Semenov Institute of Chemical Physics RAS, Kosygina 4, Moscow 119991, Russia
e-mail: suzdalev@chph.ras.ru

pearance of magnetization without superparamagnetic behavior and were observed in isolated ferrihydrate nanoclusters ($d \sim$ 1–2 nm) in porous materials and strong interacting α-, γ-Fe$_2$O$_3$ nanoclusters ($d \sim$ 20–50 nm). The phenomenon can be interpreted in the frame of thermodynamics models that take into account surface pressure and surface tension of nanoclusters with $d < 5$ nm or influence of defects in nanostructures with $d \sim$ 20–50 nm. A thermodynamic model for magnetic phase transitions in nanocluster has been proposed in [1].

A cluster is assumed to exhibit a pronounced magnetostriction effect. The Curie point of a cluster depends on its volume and can be expressed by

$$T_c = T_0 \left(1 + \beta \frac{V - V_0}{V_0} \right) \tag{1}$$

where β is the magnetostriction constant of a substance, V and V_0 are molecular volume of a substance in the magnetically ordered and paramagnetic states under action of surface tension. The temperature dependence of magnetization is given by the equation

$$T / T_{co} = m \left(\frac{1}{3} \gamma m^2 + 1 \right) \left(\frac{1}{2} \ln \frac{1 + m}{1 - m} \right)^{-1}, \tag{2}$$

where $T_{co} = T_0 (1 - \beta \eta P)$ is the change of T_0 due to pressure $P = 2\alpha/R$, $\gamma = 3/2 N k T_0 \eta \beta^2 (1 - \beta \eta P)^{-1}$ (η is compressibility, α is the surface tension constant). For $\gamma < 1$, the phase transition corresponds to the second-order transition and for $\gamma \geq 1$ to the first-order magnetic transition. Representing $\gamma = T_{cc}/T_{co}$, the critical cluster size for the appearance of the first order magnetic phase transition is given by the expression

$$R_{CR} = \frac{2\alpha\beta\eta}{1 - T_{cc}/T_{co}}, \tag{3}$$

where $T_{cc} = 3/2 N k T_0^2 \eta \beta^2$. At $R \leq R_{CR}$ and $\gamma \geq 1$, the cluster passes from the magnetically ordered to the paramagnetic state via first-order magnetic phase transition. For larger cluster, only the second-order magnetic phase transition takes place (Langeven type). The latter case can be also interpreted in terms of superparamagnetism.

For the explanation of the first-order magnetic phase transition in α-, γ-Fe$_2$O$_3$ nanosystem the thermodynamic model taking into account the influence of cluster defect density was developed [2]. By this model

$$T_c = T_{co} (1 + \beta c_v), \tag{4}$$

where $\beta < 0$ is constant, T_c, T_{co} are the Curie or Neel points for defect and perfect clusters, c_v is concentration of defects in cluster. Concentration of defects in clusters reveals maximum value for the cluster sizes $d \sim$ 30–50 nm. For $d \sim$ 3–250 nm the calculation of temperature dependence of magnetization allows to obtain the second-to-first-order change of the character of magnetic phase transition whereby the second order is observed in the range $d \sim$ 100–250 nm, the first order at $d \sim$ 6–50 nm and again second order at $d \sim$ 3 nm. The thermodynamic potential of the system may be expressed by

$$G = \Delta \mu n_v + \alpha S(n_v) - 1/2 N_0 k T_{co} (1 + \beta c_v) m^2$$
$$+ N K T \{ 1/2 \ln [(1 - m^2/4)] + m \ln [(1 + m)/(1 - m)] \} \tag{5}$$

 Springer

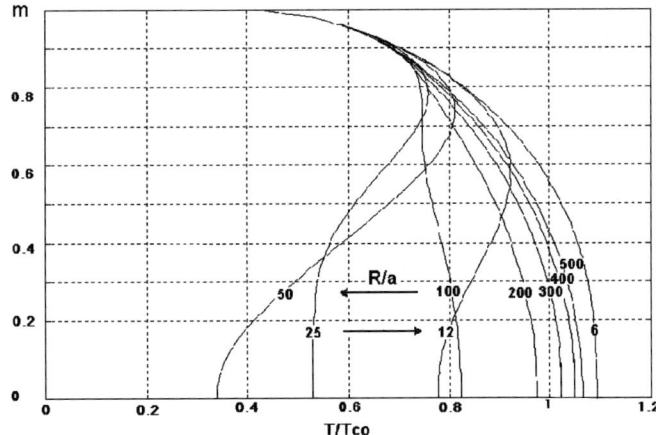

Figure 1 Temperature dependance of calculated cluster magnetization at different R/a.

where $\Delta\mu = \Delta\mu^0 + kT \ln(N_s n_v/N_v n_s)$ is the difference of chemical potential for defects situated inside and outside the cluster, N_v, N_s – numbers of atom in cluster volume and surface, respectively, n_v, n_s – concentrations in cluster volume and surface, respectively, α – the surface tension on cluster–media interface, S – cluster surface, N – number of spins in cluster, $m = M/M_0$ – relative magnetic moment of cluster, M – magnetic moment of cluster. If the conditions $dG/dn_v = 0$ and $dG/dm = 0$ are simultaneously fulfilled, the equilibrium may be expressed by the equation connecting implicitly the magnetization and temperature

$$1 + (\beta c_0)\, q\,(R) / \left\{1 + q\,(R)\, K \exp\left[(V\delta p(R)/kT_{co} - 0,5(n\cdot\beta)m^2)\,(T_{co}/T)\right]\right\}$$
$$- (0,5/m)\ln\left[(1+m)/(1-m)\right] T/T_{co} = 0 \qquad (6)$$

where $q(R) = [(1 + r/R)^3 - 1] = [(1 + (r/a)(a/R))^3 - 1]$, $\delta p(R) = 2\alpha/R$, $n = N/N_v$ is the concentration of spins in cluster, a is the lattice constant, r is the thickness of intercluster media surrounding cluster the calculation of the magnetization curve $m(T/T_{co})$. The following parameters are used $\beta c_0 = -50$, $k = \exp(-\Delta\mu^0/kT_{co}) = 0.01$, $n\beta = -3$, $r/a = 1$, $2\alpha V/akT_{co} = 50$ (Figure 1). Temperature dependence of magnetization for $R = 500a - 6a$ shows the character of magnetic phase transition changes notably from the smooth curves typical for of second-order magnetic ($R/a = 500, 400, 300, 200$) to the jump like curves characteristic of the first-order magnetic phase transition (R/a 100, 50, 25, 12) and back to second-order magnetic phase transition for ($R/a = 6$).

While decreasing nanocluster size the decrease of the Neel temperature was obtained in nanostructures fabricated by high pressure with shear action. This technique generates nanostructures with nanocrystallite sizes up to 10 nm and high density of defects. The mixture of ~5 wt.% α-,γ-Fe_2O_3 clusters ($d \sim 20$–50 nm) with ~95 wt.% metallic Eu was treated by pressure 2 GPa and shear 240°. Room temperature Mössbauer spectra [57]Fe of initial sample and sample subjected to shear loading are shown in Figure 2. The absence of hyperfine structure (HFS) lines in Figure 2b supports the idea that shear stress under high pressure induces the decrease

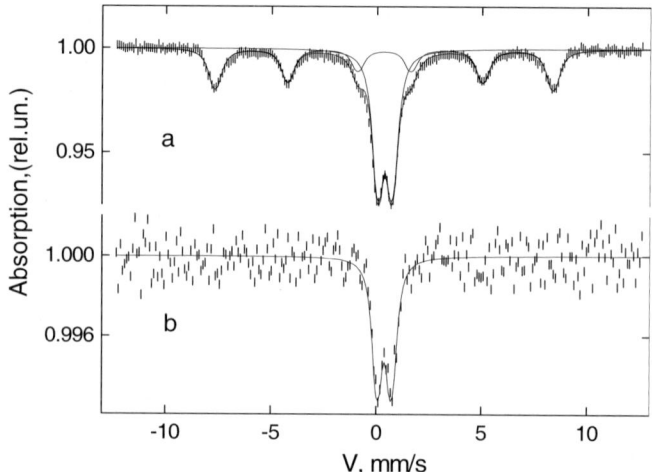

Figure 2 Room temperature Mössbauer spectra ^{57}Fe of nanostructure α-, γ-Fe$_2$O$_3$ +Eu metal: **a** – initial, **b** – the same after action of high pressure with shear.

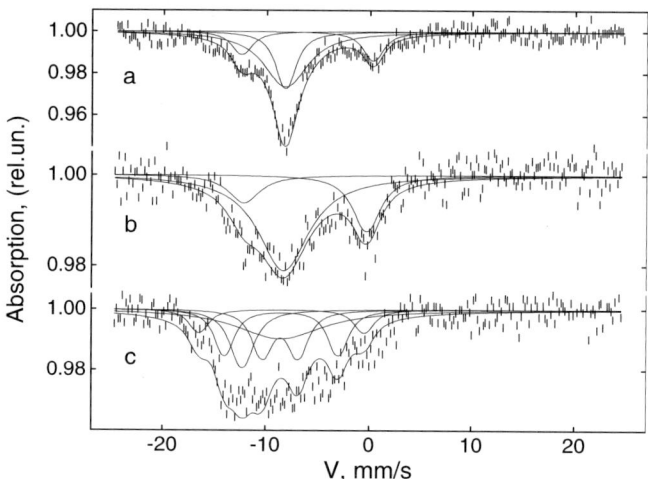

Figure 3 Mössbauer spectra at 90 K of nanocomposites containing metallic ^{151}Eu after action of high pressure with shear: **a** nanostructured Eu metal, **b** nanostructured Eu metal +α-, γ-Fe$_2$O$_3$, **c** nanostructured Eu metal+adamantane.

of cluster size accompanied by the decrease of the Curie temperature of magnetic phase transition.

On the contrary, for metallic Eu and Eu-based composites an increase of the Neel points was typical. The action of high pressure with shear was studied in three systems: Metal Eu; α-, γ-Fe$_2$O$_3$ nanoclusters (about 5 wt.%)+ metal Eu; and adamantane (about 10 wt.%) + metal Eu. The bulk metal Eu possesses the first order magnetic phase transition at 80 K. Mössbauer spectra at $T = 90$ K of ^{151}Eu are shown in Figure 3.

 Springer

One can see that for metal Eu the magnetic order and HFS lines disappear at $T = 90$ K but the broadening of paramagnetic line is still present. More pronounced line broadening effect is observed for the systems of α-, γ-Fe_2O_3 nanoclusters + metal Eu and of adamantane (about 10 wt.%) + metal Eu. In the latter case even resolved HFS lines are visible. All these results can be interpreted in terms of shear stress under high pressure induced increase of Neel temperature in metal Eu accompanied by the appearance of the second order magnetic phase transitions. For the cluster size ranging 20–50 nm there exists maximum of defect density causing first-to-second-order change of magnetic phase transitions. Thus, high pressure with shear action seems to provide the decrease of metal Eu magnetic domain less than 20–50 nm and the first-to-second-order change of magnetic phase transition accompanied by increasing of Neel points. The embedding of α-, γ-Fe_2O_3 nanoclusters and adamantane into Eu matrix permits to distribute a shear stress over all the composite volume and to conserve generated defects and stresses after the release of pressure. Thus, the severe plastic deformation allows nanocluster and nanostructure materials to change the character of magnetic phase transitions and to decrease or to increase the critical temperature of magnetic phase transitions.

Acknowledgments This work is supported by RFBR (03-03-32029) and INTAS (01-204).

References

1. Suzdalev, I.P., Buravtsev, V.N., Imshennik, V.K., Maksimov, Yu.V., Matveev, V.V., Volynskaya, A.V., Trautwein, A.X., Winkler, H.: Z. Phys. D **37**, 55 (1996)
2. Suzdalev, I.P., Buravtsev, V.N., Imshennik, V.K., Maksimov, Yu.V.: Scr. Mater. **44**, 1937 (2001)

Hyperfine Interactions (2005) 164

Author Index to Volume 164 (2005)